Cambridge Primary

Science

Second Edition

Learner's Book 6

Series editors:
Judith Amery
Rosemary Feasey

Series authors:

Rosemary Feasey
Deborah Herridge
Helen Lewis
Tara Lievesley
Andrea Mapplebeck
Hellen Ward

T0265945

Boost

HODDER
EDUCATION
AN HACHETTE UK COMPANY

Cambridge International copyright material in this publication is reproduced under licence and remains the intellectual property of Cambridge Assessment International Education.

Registered Cambridge International Schools benefit from high-quality programmes, assessments and a wide range of support so that teachers can effectively deliver Cambridge Primary. Visit www.cambridgeinternational.org/primary to find out more.

Third-party websites and resources referred to in this publication have not been endorsed by Cambridge Assessment International Education.

The audio files are free to download at www.hoddereducation.com/cambridgeextras

Acknowledgements
The Publishers would like to thank the following for permission to reproduce copyright material.

Every effort has been made to trace or contact all copyright holders, but if any have been inadvertently overlooked, the Publishers will be pleased to make the necessary arrangements at the first opportunity.

Text acknowledgements
p. 29 'United Nations World Population Ageing Report, 2017 Highlights'. https://www.un.org/en/development/desa/population/publications/pdf/ageing/WPA2017_Highlights.pdf. Attribution 3.0 IGO (CC BY 3.0 IGO), https://creativecommons.org/licenses/by/3.0/igo/; **p. 30** *tr* photo/adaption © L'Oreal Thailand For Women In Science 2018. **p. 112** *cl* illustration © Painting by Richard Bizley bizleyart.com.

Photo acknowledgements
p. 5 *cr* © Anita P Peppers/Adobe Stock Photo; **p. 5** *cr*, **p. 101** *cr* © John Anderson/Adobe Stock Photo; **p. 8** *tr* © Wckiw/Adobe Stock Photo; **p. 9** *br*, **138** *tr* © Design Cells/Adobe Stock Photo; **p. 11** *tr* © Granger Historical Picture Archive/Alamy Stock Photo; **p. 11** *cl* © Art Directors & TRIP/Alamy Stock Photo; **p. 11** *cr* © Pictorial Press Ltd/Alamy Stock Photo; **p. 11** *cl* © Science History Images/Alamy Stock Photo; **p. 20** *cc* © Carrot Eater/Shutterstock.com; **p. 24** *tr* © Rokas/Adobe Stock Photo; **p. 25** *tr* © Thananit/Adobe Stock Photo; **p. 26** *tr* © Valentina R/Adobe Stock Photo; **p. 26** *cc* © Sonne Fleckl/Adobe Stock Photo; **p. 26** *cc* © Unjiko/Shutterstock.com; **p. 26** *cl* © Maria/Adobe Stock Photo; **p. 26** *cc* © Maxter Design/Adobe Stock Photo; **p. 26** *cc* © Marleah/Adobe Stock Photo; **p. 26** *cl* © Samuel B/Adobe Stock Photo; **p. 26** *cc* © Mechanik/Adobe Stock Photo; **p. 26** *cc* © Oly 5/Adobe Stock Photo; **p. 26** *cl* © Lurs/Adobe Stock Photo; **p. 26** *cc* © Anthony Brown/Adobe Stock Photo; **p. 26** *cc* © Foto Zick/Adobe Stock Photo; **p. 27** *tr* © Prostock-studi/Adobe Stock Photo; **p. 32** *cr*, *br* © Hachette UK; **p. 32** *br* © Hachette UK; **p. 34** *cr*, *br* © Dave/Adobe Stock Photo; **p. 34** *bl* © Stock Photo Mania/Adobe Stock Photo; **p. 34** *br* © Photocech/Adobe Stock Photo; **p. 38** *tr* © CW Images/Alamy Stock Photo; **p. 40** *tr* © Aleksandar kamasi/Adobe Stock Photo; **p. 43** *tl* © Stone 36/Adobe Stock Photo; **p. 43** *tc*, **p. 49** *cr* © Tab/Adobe Stock Photo; **p. 43** *tr* © Franz 12/Adobe Stock Photo; **p. 43** *cl*, **p. 49** *cr* © Africa Studio/Adobe Stock Photo; **p. 43** *cc* © Aquar/Adobe Stock Photo; **p. 43** *cr* © Irina Burakova/Adobe Stock Photo; **p. 43** *br* © Fifth Dimension/Adobe Stock Photo; **p. 48** *cl* © Avant Garde/Adobe Stock Photo; **p. 48** *cc* © Hachette UK; **p. 48** *cr* © Volod 2943/Adobe Stock Photo; **p. 49** *cr*, **p. 46** *tr* © Zig Koch/Adobe Stock Photo; **p. 49** *cc* © Photocreo Bednarek/Adobe Stock Photo; **p. 49** *cc*, **p. 54** *cc* © Schankz/Adobe Stock Photo; **p. 51** *br* © Gregory Johnston/Adobe Stock Photo; **p. 54** *cc* © New Africa/Adobe Stock Photo; **p. 54** *cr*, *cr* © New Africa/Adobe Stock Photo; **p. 54** *cc* © Hachette UK; **p. 54** *cr* © Aquar/Adobe Stock Photo; **p. 54** *bc* © Muhammad Sainudin/Adobe Stock Photo; **p. 55** *bl* ©Hachette UK; **p. 56** © Evgeniy/Adobe Stock Photo; **p. 61** *tr* © Mark Ka/Adobe Stock Photo; **p. 62** *tr* © Praphab 144/Adobe Stock Photo; **p. 62** *br* © D R 3D/Adobe Stock Photo; **p. 63** *br* © Veronika Synenko/123rf; **p. 64** *tr* © Greg Meland/Adobe Stock Photo; **p. 64** *br* © Diana Taliun/Adobe Stock Photo; **p. 73** *tr* © Askolds/Adobe Stock Photo; **p. 79** *tr* © Hachette UK; **p. 81** *tc* © Gabees/Adobe Stock Photo; **p. 81** *tr* © Oleksandrum/Adobe Stock Photo; **p. 85** *tc* © Maya Kruchancova/Adobe Stock Photo; **p. 93** *cr* © Hachette UK; **p. 94** *tr* © Gail Johnson - Fotolia; **p. 97** *bl* © Cliff Hide News/Alamy Stock Photo; **p. 98** *cl* © Sichkarenko com/Adobe Stock Photo; **p. 98** *cc* © Willyam/Adobe Stock Photo; **p. 98** *cc* © KTS Design/Adobe Stock Photo; **p. 98** *cr* © Герман Бикаев/Adobe Stock Photo; **p. 98** *bl* © 1xpert/Adobe Stock Photo; **p. 98** *br* © Givaga/Adobe Stock Photo; **p. 100** *bc* © Hachette UK; **p. 105** *cl* © Gavin Newman/Alamy Stock Photo; **p. 106** *tr* © Antonio Violi/Alamy Stock Photo; **p. 107** *tc* © Michal 812/Adobe Stock Photo; **p. 107** © Zelenka 68/Adobe Stock Photo; **p. 110** © Andrea Cerri Ferrari/Adobe Stock Photo; **p. 111** *tr* © Brooke Becker - Fotolia.com; **p. 114** *cr* © Sci/Adobe Stock Photo; **p. 115** *tr* © SINCLAIR STAMMERS/SCIENCE PHOTO LIBRARY; **p. 115** *cr* © Smuki/Adobe Stock Photo; **p. 116** *cl* © Tami Freed/123rf; **p. 116** *cr* © Elena/Adobe Stock Photo; **p. 117** *br* © Hachette UK; **p. 118** *tc* © Universal Images Group North America LLC/Alamy Stock Photo; **p. 118** *tr* © Gustavo Ramirez/Shutterstock.com; **p. 119** *tr* © Vladislav 333222/Adobe Stock Photo; **p. 119** *cc* © Adwo/Adobe Stock Photo; **p. 121** *tr* © 994 Yellow/Adobe Stock Photo; **p. 128** *cr*, **140** *cr* © Christos Georghiou/Shutterstock.com; **p. 128** *br* © Pyty/Adobe Stock Photo.com; **p. 129** *tr* © David Carillet/Adobe Stock Photo; **p. 131** *tr* © Exsodus/Adobe Stock Photo; **p. 131** *cl* © Vchalup/Adobe Stock Photo; **p. 132** *br* © Taff Pixture/Adobe Stock Photo; **p. 133** *br* © Alejomiranda/Adobe Stock Photo; **p. 134** *cl* © Paulista/Adobe Stock Photo; **p. 135** *cr* © NASA/JPL-Caltech; **p. 136** *tr* © Claudia Weinmann/Alamy Stock Photo; **p. 136** *cr*, **p. 143** *bc* © Skyelar/Adobe Stock Photo.

t = top, *b* = bottom, *l* = left, *r* = right, *c* = centre

Hachette UK's policy is to use papers that are natural, renewable and recyclable products and made from wood grown in well-managed forests and other controlled sources. The logging and manufacturing processes are expected to conform to the environmental regulations of the country of origin.

Orders: please contact Hachette UK Distribution, Hely Hutchinson Centre, Milton Road, Didcot, Oxfordshire, OX11 7HH. Telephone: +44 (0)1235 827827. Email education@hachette.co.uk. Lines are open from 9 a.m. to 5 p.m., Monday to Saturday, with a 24-hour message answering service. You can also order through our website: www.hoddereducation.com

© Rosemary Feasey, Deborah Herridge, Helen Lewis, Tara Lievesley, Andrea Mapplebeck, Hellen Ward 2021

First published in 2017
This edition published in 2021 by
Hodder Education,
An Hachette UK Company
Carmelite House
50 Victoria Embankment
London EC4Y 0DZ

www.hoddereducation.com

Impression number   10 9 8 7 6 5 4 3 2
Year   2025 2024 2023 2022 2021

All rights reserved. Apart from any use permitted under UK copyright law, no part of this publication may be reproduced or transmitted in any form or by any means, electronic or mechanical, including photocopying and recording, or held within any information storage and retrieval system, without permission in writing from the publisher or under licence from the Copyright Licensing Agency Limited. Further details of such licences (for reprographic reproduction) may be obtained from the Copyright Licensing Agency Limited, www.cla.co.uk

Cover illustration by Lisa Hunt, The Bright Agency

Illustrations by James Hearne, Natalie and Tamsin Hinrichsen, Stéphan Theron, Steve Evans, Vian Oelofsen

Typeset in FS Albert 15/17 by IO Publishing CC

Printed in Italy

A catalogue record for this title is available from the British Library.

ISBN: 9781398301771

Contents

How to use this book

This book will help you learn about Science in different ways.

Talk about what you remember or know about a topic.

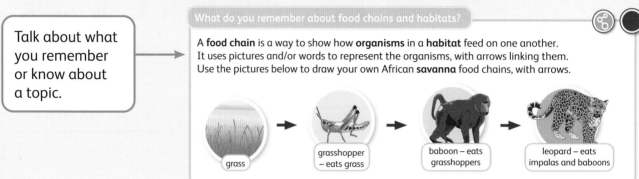

What do you remember about food chains and habitats?

A **food chain** is a way to show how **organisms** in a **habitat** feed on one another. It uses pictures and/or words to represent the organisms, with arrows linking them. Use the pictures below to draw your own African **savanna** food chains, with arrows.

grass → grasshopper – eats grass → baboon – eats grasshoppers → leopard – eats impalas and baboons

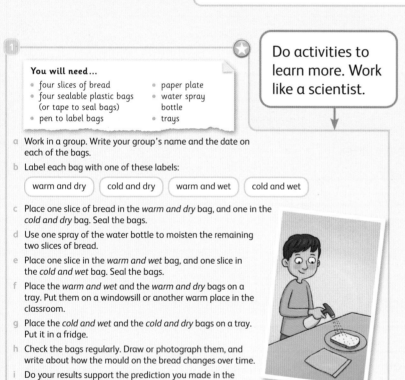

You will need...
- four slices of bread
- four sealable plastic bags (or tape to seal bags)
- pen to label bags
- paper plate
- water spray bottle
- trays

a Work in a group. Write your group's name and the date on each of the bags.

b Label each bag with one of these labels:

| warm and dry | cold and dry | warm and wet | cold and wet |

c Place one slice of bread in the *warm and dry* bag, and one in the *cold and dry* bag. Seal the bags.

d Use one spray of the water bottle to moisten the remaining two slices of bread.

e Place one slice in the *warm and wet* bag, and one slice in the *cold and wet* bag. Seal the bags.

f Place the *warm and wet* and the *warm and dry* bags on a tray. Put them on a windowsill or another warm place in the classroom.

g Place the *cold and wet* and the *cold and dry* bags on a tray. Put it in a fridge.

h Check the bags regularly. Draw or photograph them, and write about how the mould on the bread changes over time.

i Do your results support the prediction you made in the *Let's talk* activity?

Do activities to learn more. Work like a scientist.

Think like a scientist!

Planets are **rotating** (spinning) all the time. They spin around an imaginary line called an **axis**. The axis of the Earth is tilted at 23.5 degrees.

One complete **rotation** on the axis of the planet is called a **day**.

On Earth it takes 24 hours to make one complete rotation. This is the length of a day on Earth.

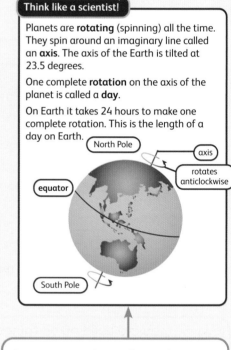

North Pole | axis | rotates anticlockwise | equator | South Pole

Learn new ideas about Science.

Let's talk

Discuss whether you or anyone in your family has ever been sick. What do you think caused these diseases?

Talk about your ideas.

Learn about interesting facts and information.

Did you know?

The total mass of Earth's atmosphere is about 5.7 quadrillion tons. That's about the same as 570,000,000,000,000 adult Indian elephants!

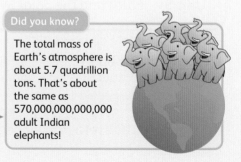

Scientific word
vertebrates

Understand new words. The *Scientific dictionary* at the back of the book can also help you.

Challenge yourself!
If you lick your finger and move it in the air, it feels colder. Why do you think this happens?

What can you do?

You have learnt about human reproduction. You can:

✔ name the parts of the male and female reproductive systems.

✔ describe the different life stages that humans go through.

✔ describe the physical changes that take place during puberty in males and females.

✔ explain why it is important for all animals to reproduce.

Try something new.

Find out how much you have learnt and what you can do.

Learn how we all use Science every day in our lives.

Model icon

Shows you are using a mental or physical model of something in the real world.

Star icon

Shows you need to think and work like a scientist.

Link icon

Shows you are learning things that link to another subject.

Progression icon

Shows you are building on things you learnt in other stages.

Audio icon

Indicates that the content is available as audio. All audio is available to download for free from www.hoddereducation.com/cambridgeextras.

There are online resources at boost-learning.com.

Science in context

Bioluminescence

People are not able to create their own light. We have to rely on light sources, such as the Sun and light bulbs, to create light for us. However, there are many animals that can create light by themselves. Scientists study these animals to see what we can learn from them to help humans develop better ways of lighting up the world.

The ability of an animal to create its own light is known as **bioluminescence**. A well-known bioluminescent animal is the firefly.

Scientists have also found that under the right conditions bubbles can give off light. Bubbles do this through a process known as **sonoluminescence**. Bubbles are little pockets of gas. When sound waves pass through them, they squash the bubbles. The bubbles then release energy in a fantastic burst of heat and light. This happens in nature when snapping shrimps clamp their claws shut.

The findings of these scientists might yield cleaner and more efficient sources of energy for us in the future. This would save fuel, and help to protect our planet from climate change.

firefly

snapping shrimp

Scientific words
bioluminescence
sonoluminescence
luminescent

Scientists have been able to make materials that are **luminescent** (give off light).

You are going to design an item of luminescent clothing for someone to wear.

a Think about a specific person who might find it useful to wear an item of clothing that gives off light. Will it give off light all the time, or will they be able to turn it on and off when they need to?

b Design your item of clothing. Produce an annotated poster showing where the luminescent material is and how it helps the person for whom you have designed it.

Let's talk

a Discuss the words *bioluminescence* and *sonoluminescence* with a partner.

b What do you think *bio-*, *sono-* and *luminescence* mean?

c Discuss your ideas with another group.

Work safely!
ONLY an adult should carry out the activities on this page. Stand well back while observing.

Always work safely.

Be a scientist

We are thinking and working like scientists when …

We describe how models can be used to show scientific ideas and what happens in science.

We present data using bar charts, dot plots, line graphs and scatter graphs.

We use scientific understanding to interpret results, and data and observations to make conclusions.

We use scientific knowledge to make predictions and check how accurate they were.

We decide what to measure and if we must repeat readings to get reliable results.

We ask questions and know which scientific enquiry activities to use to answer them.

We use different sources to research answers to questions.

We take accurate measurements and record observations and data.

We identify and reduce risks so that we can work safely.

We choose equipment to carry out scientific investigations.

We say how to improve investigations and explain why.

We classify things and use keys.

We plan fair tests where we identify and control variables.

We describe patterns in data and identify any unexpected readings.

We choose which scientific enquiries to use to answer scientific questions.

Scientific enquiry activities

When scientists work, they ask lots of questions, and think about how to answer their questions. There are different ways that scientists can answer their questions. These are called scientific enquiries. There are five different types of scientific enquiry.

1 Research

You answer the question by finding information, for example from books, video clips, the internet, leaflets and people. Example question that you can answer through research: *What solar-powered lighting solutions are scientists developing?*

2 Fair test

You carry out a test where you change something, measure something and keep some things fair or the same. Example question that you can answer through a fair test: *How does the temperature of the air affect evaporation?*

3 Observe over time

You use one or more of your senses to observe what happens over time. Example question that you can answer by observing over time: *What happens to water when you leave it outside?*

4 Identify and classify

You sort, group and name things. Example question that you can answer by identifying and classifying: *What are the properties of these rocks?*

5 Pattern seeking

You look for patterns and connections in data. Example question that you can answer by pattern seeking: *What is the relationship between mass and weight?*

Systems and diseases

What are your body's organs?

What do you remember about your body's organs?

An **organ** is a structure (part) in the human body that carries out a certain function (job).

Let's talk

Discuss these questions with a partner:

a What other external organs can you name?

b What is the function of these organs?

c Link three words from the *Scientific words* box below into a sentence. See how many different sentences you can make, using these words.

Think like a scientist!

Some organs are on the outside of the body. We call these **external organs**.

The ears are external organs. Their function is to detect and transmit sound waves to the brain.

Many organs are inside the body. We call these **internal organs**.

The kidneys are internal organs. Their function is to remove **waste** products from the blood.

1

Match the organ in each body picture with its correct name and function. Write the letter of the picture, the name of the organ and its function.

Names of organs:

(stomach)　(heart)　(intestines)　(brain)　(lungs)

Functions:

- controls the body
- takes in **oxygen**
- extracts (removes) **nutrients** from food
- pumps blood
- breaks down food

Scientific words

organ
external organs
internal organs
waste
stomach
heart
intestines
brain
lungs
oxygen
nutrients

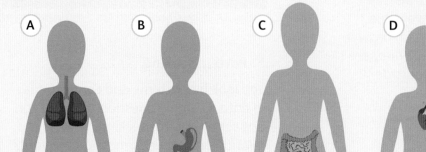

A　　B　　C　　D

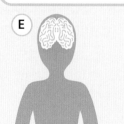

E

The circulatory system

Think like a scientist!

When two or more organs in the body work together to carry out a task, it is called a body system. An important body system is the **circulatory system**, which includes the heart and **blood vessels**. They work together to **circulate** (move) **blood** around the body.

The heart pumps blood through the blood vessels. Blood vessels are tubes that form a complex network that covers every part of the body.

Blood delivers oxygen and nutrients to the **cells** of your body. Cells are the basic units that make up your body. Blood also carries away waste products.

Each time your heart beats, you can feel it as a **pulse** on some parts of your body. The pulse you feel is the blood rushing through your blood vessels.

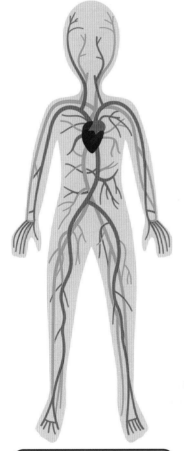

human circulatory system

Scientific words

circulatory system	blood vessels
circulate	blood
cells	pulse

1

a Use your first two fingers to feel for your pulse. Do not use your thumb.

b Feel for your pulse on the side of your neck, just under your jaw. Or feel for it on your wrist, on the same side as your thumb.

c Start by touching the skin gently. Then gradually press a little harder.

2

a Look at your hands and wrists. Can you see any blood vessels?

b Draw what you see.

c Use what you have learnt about the circulatory system to add labels to your drawing.

Did you know?

Your blood vessels would go more than twice around the Earth if you could stretch them into a long line – that is about 100 000 kilometres!

The heart

Think like a scientist!

Your heart is part of your circulatory system. It is a strong muscle that is inside your chest, a little to the left of the centre. Make a fist with your hand. This is about the size of your heart. Your heart is always beating and pumping blood around your body. It contracts and relaxes in order to do this. The process is happening all the time, but there are six stages:

1 Your heart pumps blood to your lungs.

2 In the lungs, the blood picks up oxygen from the air that you have breathed in.

3 The oxygen-rich blood travels back to your heart.

4 The heart gives the blood a second push. This time, the blood travels to all the other parts of your body.

5 The blood gives oxygen to the rest of the body.

6 The blood, which is now low in oxygen, travels back to the heart. The process begins again.

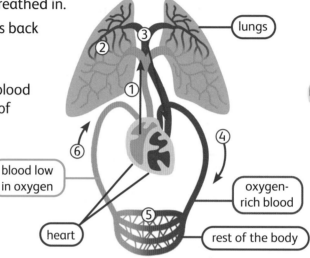

1

You will need...
- timer or stopwatch

Your **heart rate** is a measure of how many times your heart beats (pumps blood) in 1 minute. Measure your heartbeat:

a Feel your pulse. Count how many times you can feel your heartbeat in 20 seconds.

b Calculate your heart rate (number of times you felt your heartbeat in 20 seconds x 3).

c Compare your heart rate with others. Does everyone have the same heart rate?

2

a Jog for 2 minutes on the spot. Then take your pulse and calculate your heart rate.

b Compare your heart rate after exercise to your heart rate before exercise (from Activity 1). Discuss with a partner what you notice. Share your ideas with the class.

Why do you think your heart rate after exercise is different to your heart rate before exercise?

Scientific word
heart rate

Science in context

How ideas about the circulatory system have changed

Galen was a Greek doctor in the 2nd Century who tended to the Emperors in Rome. He thought that the body had two one-way systems for blood. He said that the first system had brightly coloured blood with oxygen in it. It came from the heart and gave the body air. The second system had blue blood without oxygen in it. It came from the liver and gave the body food.

The first person to question Galen's beliefs was the 13th-Century Syrian physician, Ibn al-Nafīs. Both Ibn al-Nafis and Michael Servetus, a 16th-Century Spanish physician, believed that there was a circulatory system that took blood from the heart to the lungs and back. But they thought it did not go to any other parts of the body.

The discovery that the heart pumped blood everywhere in the body was only made in the 17th Century by William Harvey, doctor to the King of England. As a scientist, Harvey asked many questions, carried out many experiments, and developed many theories. One main experiment was on how much blood flowed through the heart. He published his ideas in 1628, but many people refused to believe him. Eventually the world came to see that he was correct. He changed the way we think about the circulatory system forever.

Today we know that the heart pumps blood to every part of the body, around one single system of blood vessels in the human body.

1

In a small group, choose one of the activities below to explain how ideas about the circulatory system changed. Whichever activity you do, make sure that you use at least ten key words. Look back at the *Scientific word* boxes to help you decide which words to use. Do further research if you need to.

- Write and act out a play.
- Produce a cartoon strip with at least eight images.
- Write and perform a song.

Let's talk

With a partner, discuss the different ideas people believed about the circulatory system and how they have changed over time.

Heart rates in different vertebrates

Think like a scientist!

Animals that have a backbone are called **vertebrates**. Humans are one type of vertebrate. Across the animal kingdom, there are lots of different vertebrates. Many have circulatory systems similar to humans. They all have different heart rates.

Scientific word

vertebrates

Let's talk

With your partner, discuss the following questions:

a Do you think all animals have the same heart rate? What are your reasons for your ideas?

b Does the size of an animal make a difference to its heart rate?

c Order the animals from the fastest to the slowest heart rate. Explain why you have put them in this order.

A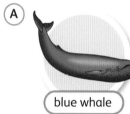
blue whale

B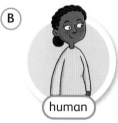
human

C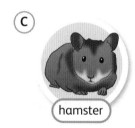
hamster

D
horse

E
domestic cat

F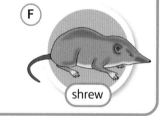
shrew

1

a Do research to find out the size of each animal in the *Let's talk* activity, and how often each animal's heart beats every minute.

b Use the data you collect about the heart rates of the animals to plot a scatter graph. Put the size of the animal on the *y*-axis and the heart beats per minute of the animal on the *x*-axis. What patterns do you notice about the size of the animal and how often its heart beats?

c Try to explain the patterns you identify.

d Check that you put the animals in the correct order for the *Let's talk* activity.

e Look at your data and predict the heart rate for a lion and a hummingbird. Explain why you have chosen these values.

lion

hummingbird

Blood vessels

Think like a scientist!

Blood vessels are part of the circulatory system. They are tubes that carry blood around your body. There are three main types of blood vessels:

- **Arteries** carry blood away from the heart, to the capillaries.
- **Capillaries** are tiny blood vessels. They deliver water, oxygen and nutrients to the body cells. They also carry away waste products.
- **Veins** carry blood away from the capillaries and back to the heart.

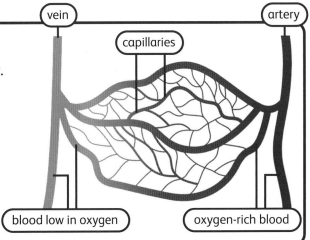

vein artery capillaries blood low in oxygen oxygen-rich blood

1

Some Class 6 learners designed this working model of the circulatory system.

a Design your own model of the circulatory system. Include these parts:

heart   blood   artery   capillaries   vein

b List what you need to make your model.

c Make the model you designed.

d Photograph your model. Add labels to your photograph and write a description of how it shows the circulatory system working.

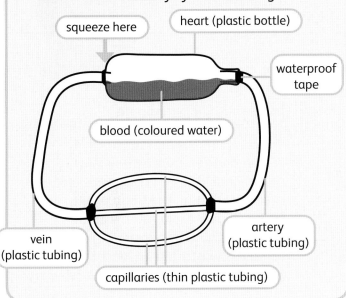

squeeze here   heart (plastic bottle)   waterproof tape   blood (coloured water)   vein (plastic tubing)   artery (plastic tubing)   capillaries (thin plastic tubing)

Scientific words

arteries   capillaries   veins

Did you know?

There is no such thing as blue blood! All blood in the human body is red. Oxygen-rich bright red blood flows through arteries and capillaries.

Dark red blood flows back to the heart through veins. It is dark red because it is low in oxygen, having given up the oxygen to the different body cells. People think veins have blue blood as veins close to the skin surface look blue.

2

Think about your model from Activity 1:

a Which parts worked well in helping to explain the circulatory system?

b What parts did not work well to explain the system? Why?

c How could you extend your model to show:
- how blood transports oxygen, nutrients and waste?
- how the system links to other internal organs in the body?

Investigate the circulatory system

Let's talk

When you stop cycling, does it take your heart a while to stop beating fast?

Discuss Guss's question with a partner. Share your ideas with another pair.

1

Work in a group. Plan an investigation to find out: *How quickly does your heartbeat recover (go back to normal) after exercise?*

a How can you find out if this is the same for everyone?

b How many people will you test?

c What will you measure?

d How will you make sure that the data you collect is reliable?

Challenge yourself!

Do research to find out different ways you can keep your heart healthy.

2

a Carry out the investigation you planned in Activity 1.

b Present your results in a table. Include a way to look at repeated results (such as calculating averages).

c What conclusion can you draw from your results?

d Make a further prediction based on your results.

3

Sanchia has some questions about blood and circulation.

What is blood made from?

Is blood made in the heart?

Is the function of the heart to clean blood?

Do arteries contain clean blood and veins dirty blood?

a Do research to answer Sanchia's questions. Share what you find out with a partner.

b With a partner, create a role play to show Sanchia what happens as blood circulates around the body. Make sure your role play explains how blood transports nutrients, including oxygen, as well as waste. Use at least ten key words from pages 9–13 in your role play.

The respiratory system

Think like a scientist!

The **respiratory system** is made up of organs that work together to allow **gas exchange** between the atmosphere (air around you) and the body. Gas exchange happens when oxygen and carbon dioxide move in opposite directions across a surface. In humans, this happens in the lungs, as well as in every cell. In humans, we call this gas exchange **breathing**. It helps us get oxygen from the atmosphere and pass it into the blood.

In humans and many other vertebrates, there are three parts to the respiratory system:

- tubes that carry air from the atmosphere towards the lungs
- lungs
- a muscle called the **diaphragm**.

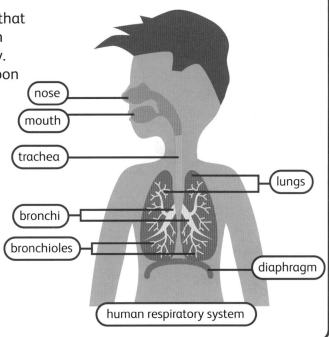

human respiratory system

1

Copy this table and do research to complete it. It will help you understand why each part of the respiratory system is important. The labels in the diagram above will help you. The trachea has been completed for you.

Scientific words

respiratory system
gas exchange
breathing
diaphragm
trachea
bronchi
bronchioles

The respiratory system			
Parts	**Part involved**	**What if this organ were missing?**	**Function of this part**
tubes	nose		
tubes	mouth		
tubes	**trachea**	no air could enter or leave the lungs	• allows air to pass in and out of the lungs • warms the air and cleans it of dirt
lungs	**bronchi**		
lungs	**bronchioles**		
diaphragm	diaphragm		

Breathing

Think like a scientist!

As you **inhale** (breathe in), your diaphragm muscle contracts (gets smaller), and moves down. Your ribs lift up and out. This gives your lungs space to fill up with air. Oxygen from the air passes into your lungs, and then into your bloodstream.

As you **exhale** (breathe out), your diaphragm muscle relaxes and moves back up. Your ribs move in. This makes your chest smaller, and pushes air out of your lungs. This includes the waste gas, carbon dioxide.

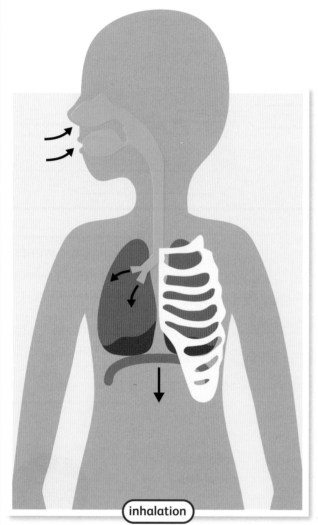

inhalation

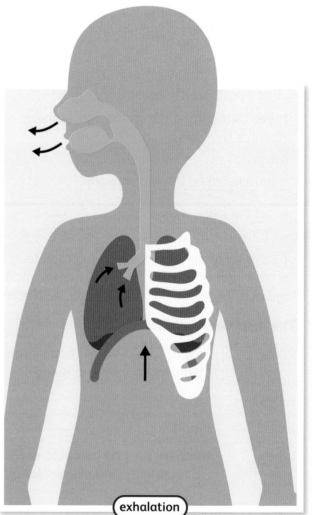

exhalation

The two-stage process of breathing is called a **complete breath**. Your **breathing rate** is the number of complete breaths you take in 1 minute.

Scientific words

inhale	exhale
inhalation	exhalation
complete breath	breathing rate

A model lung

1

You will need...
- plastic bottle
- scissors
- two balloons

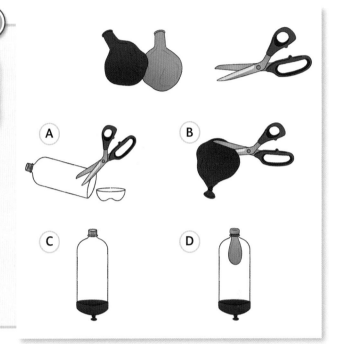

Make your own model of a lung.

A Carefully cut the bottom off the plastic bottle.

B Tie a knot in the end of one balloon. Then cut the top off.

C Stretch the balloon over the bottom of the bottle.

D Hang the other balloon inside the bottle and stretch its end over the neck of the bottle.

2

Use your model of a lung and what you have learnt so far to answer these questions.

a What is the name of the gas from air which passes into your bloodstream in the lungs?

b Explain how your body makes space for the lungs to fill with air.

c Name the gases that you breathe out.

d Explain how your body pushes air out of the lungs.

Let's talk

Discuss these questions with a partner.

a What conclusions can you draw from the results of your investigation into breathing rates?

b Predict your breathing rate during sleep. Explain your thinking.

Work safely!

Be careful with scissors. Ask your teacher to help if it is difficult to cut the bottle.

3

a Predict the answer to Pia's question below. Explain your thinking.

b Plan an investigation to answer this question.

c Carry out the investigation.

d Present your results in a line graph.

Is your breathing rate different after exercise?

Investigate breathing

Think like a scientist!

The volume of air that your lungs can hold is called your **lung capacity**.

Let's talk

How many complete breaths do you think you take in one day?

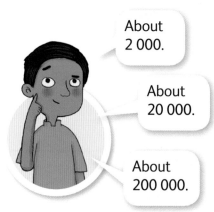

About 2 000.

About 20 000.

About 200 000.

Discuss these questions with a partner:

a Estimate the number of complete breaths you take in one day. Explain your thinking.

b How could you find out if your estimate is correct?

c What evidence will you collect?

d How will you ensure that your results are reliable?

e What equipment will you use?

Scientific word
lung capacity

18

1

a With a partner, carry out the test you discussed in the *Let's talk* activity.

b Write a short report that explains what you did and what you found out. Share your report with other pairs.

2

Predict and write down whether different people have different lung capacities. Explain.

3

You will need...
- large bowl
- large plastic bottle
- plastic tubing
- marker pen
- measuring jug

Work in a group to measure lung capacity.

a Half-fill the bowl with water.

b Fill the plastic bottle to the very top with water, and screw on the lid.

c Turn the bottle upside down and place it in the bowl of water. Take the lid off the bottle while it is under the water.

d Push one end of the tubing into the bottle and hold the other end of the tubing.

e Take a big breath. Then breathe out as much air as you can into the tube until your lungs are empty.

f All the air that you breathe out gets trapped in the bottle. Draw a line to mark the water level on the bottle.

g To measure how much air (the volume) you blew into the bottle, empty the water out the plastic bottle. Then pour in water up to the line you marked. Pour this water into a measuring jug to find its volume. This is your lung capacity.

h Record your results in a table. Check if your results support your prediction in Activity 2.

i Compare your results with others in your group. Do different people have different lung capacities?

Infectious diseases

Think like a scientist!

Our body organs and systems work together to perform important functions that keep us alive. We need to keep them healthy. To do this, we must exercise regularly, eat a **balanced diet** and protect ourselves from **infectious diseases**. Infectious diseases are one of the leading causes of death across the world.

There are different causes of infectious diseases. Some germs are too small to be seen with the eye. These germs are called **microbes**. We can only see them with a microscope, so we describe them as **microscopic**. Many microbes live in and on our bodies. They are normally harmless, and can even be helpful to us. But under certain conditions, some may cause disease.

Challenge yourself!

Micro is a prefix – it goes at the beginning of many words. It means small or very small. How many words beginning with *micro* can you think of?

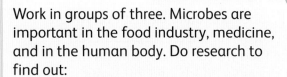

Work in groups of three. Microbes are important in the food industry, medicine, and in the human body. Do research to find out:

- how the food industry uses microbes to make yoghurt
- how microbes are important in medicine in the antibiotic, penicillin
- how microbes are important in the human digestive system.

Share your findings with others.

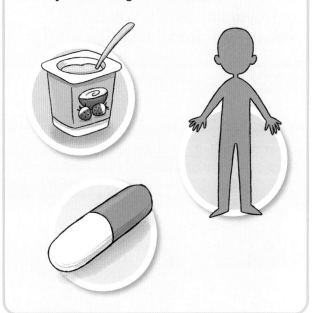

Let's talk

Discuss whether you or anyone in your family has ever been sick. What do you think caused these diseases?

Scientific words

balanced diet	infectious disease
microbes	microscopic

Microbes

Think like a scientist!

Microbes are found almost everywhere on Earth – in the air we breathe, on the food we eat, on our skin, on the surfaces we touch, and inside our bodies – in the mouth, nose and stomach.

If harmful microbes enter the body, we say the person has been **infected**. An infection does not always result in **disease**! A disease is when the infection causes damage to the organs or systems. Harmful microbes that can cause diseases include **bacteria**, **viruses**, **fungi** and **parasites**.

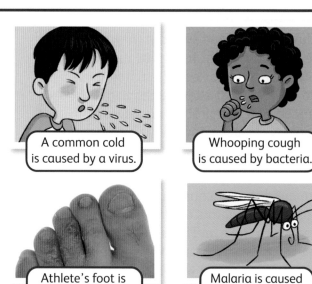

A common cold is caused by a virus.

Whooping cough is caused by bacteria.

Athlete's foot is caused by a fungus.

Malaria is caused by a parasite.

Did you know?

There are over 7.5 billion people living on Earth, and up to 50 billion bacteria in one handful of soil!

Let's talk

Discuss what you know about the different microbes.

Scientific words

infected	disease	bacteria
viruses	fungi	parasites

1

Do research to find out the order, from smallest to largest, of the different microbes.

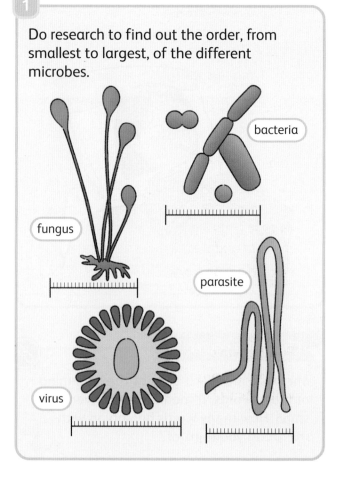

bacteria

fungus

parasite

virus

Investigate microbes

Let's talk

Mould is a type of fungus.

a Predict which of these different combinations of conditions you think will help mould grow the best on bread. Put them in order from the most mouldy to the least mouldy.

> warm and dry   cold and dry   warm and wet   cold and wet

b Explain your reasoning.

1

You will need...

- four slices of bread
- four sealable plastic bags (or tape to seal bags)
- pen to label bags
- paper plate
- water spray bottle
- trays

Work safely! ⚠

Do not taste this food.

At the end, throw the bag into a refuse bin, without opening it.

a Work in a group. Write your group's name and the date on each of the bags.

b Label each bag with one of these labels:

> warm and dry   cold and dry   warm and wet   cold and wet

c Place one slice of bread in the *warm and dry* bag, and one in the *cold and dry* bag. Seal the bags.

d Use one spray of the water bottle to moisten the remaining two slices of bread.

e Place one slice in the *warm and wet* bag, and one slice in the *cold and wet* bag. Seal the bags.

f Place the *warm and wet* and the *warm and dry* bags on a tray. Put them on a windowsill or another warm place in the classroom.

g Place the *cold and wet* and the *cold and dry* bags on a tray. Put them in a fridge.

h Check the bags regularly. Draw or photograph them, and write about how the mould on the bread changes over time.

i Do your results support the prediction you made in the *Let's talk* activity?

Control the spread of diseases

Think like a scientist!

The spreading of microbes is called **transmission**. Many harmful organisms and microbes can be transmitted from one person to another in different ways, such as through the air (by coughing and sneezing), by touch (hugging, shaking hands or touching objects), in water, on food, and by animals.

There are between 10 000 and 10 million bacteria on each of your hands! Damp hands spread 1 000 times more germs than dry hands. The number of bacteria on your fingertips doubles after you use the toilet. One person can spread almost one million bacteria in one school day.

Hand washing is one of the best **hygiene** practices to stop the spread of many diseases. Clean hands can stop bacteria from moving from one person to another and through a whole community.

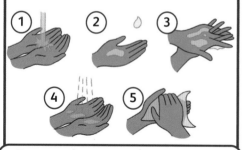

Five steps for proper handwashing: wet hands with water, rub with soap, wash hands for at least 20 seconds, rinse and dry.

Scientific words

transmission   hygiene

Did you know?

You should spend at least 20 seconds washing your hands each time. That is about as long as it takes you to sing the 'Happy Birthday' song twice! Try it the next time you wash your hands.

Let's talk

a How many different situations can you think of when you should wash your hands?

b Why is it important to use good standards of hygiene to control and stop the spread of diseases?

c How many different objects in your classroom do you think the microbes on the hand of one person could spread to in one hour?

You will need...
- hand gel mixed with brightly coloured glitter
- soap, water, paper towels

a To investigate the transmission of microbes, your teacher will squeeze some glitter hand gel onto the hands of one learner. The learner should rub the glitter gel all over the palms of their hands.

b The learner should then shake hands with three other learners in your class.

c At the end of one hour, in how many different places can you find glitter? Has it spread to all the objects you thought it would? Has it spread to more objects or fewer objects?

d As a class, use soap, water and paper towels to clean up all of the glitter you can find!

Human defence mechanisms: Investigate sneezes

Think like a scientist!

The most common way to spread infectious diseases is through particles in the air that come from a person's mouth or nose when they cough or sneeze. We can breathe in the air that contains microbes and become infected when these spread to the nose, throat and other air passages. The body produces slimy mucus to capture the microbes. It then makes us sneeze and cough to blow the microbes out of the body with the mucus. This is one way the human body has developed a **defence mechanism** to help us fight infectious diseases.

A human sneeze can travel up to 160 kilometres per hour, and there can be up to 100 000 microbes in one sneeze. When you cough, germs can travel about three metres if you do not put your hand or a handkerchief over your nose and mouth. Bacteria can live for up to 40 minutes on the surfaces on which they land.

Scientific word
defence mechanism

1

You will need...

- 'sneeze spray' bottle
- 'sneezing table' covered in plain wallpaper
- tape measures
- plastic glove
- piece of kitchen roll

a Work in a group. Use the equipment to design a test to investigate how far a sneeze will travel forward and how far it will spread outwards.

b Compare your results with others. What conclusions can you draw?

c What do your results tell you about the best way to stop diseases from spreading through the air?

2

Microbes can enter the body through:

- openings, such as the mouth and nose, and get into the stomach
- different body surfaces, such as the eyes
- breaks, cuts or insect bites on the skin.

a Do research to find out:
  - how your body defends itself if microbes enter it
  - how doctors and nurses prevent microbes from entering their bodies.

b Produce a safety leaflet to explain how we can prevent microbes from entering the body.

What is a Public Health Emergency?

Occasionally there are serious, unusual or unexpected diseases that put people throughout the world in danger. They are classified as Public Health Emergencies of International Concern (PHEIC). The concern is that these diseases can spread quickly and put other countries at risk. The countries affected by the disease may need help from other countries to treat and get rid of the disease.

There have only been a small number of diseases that have been classified as Public Health Emergencies. Three examples are the Zika virus, the Kivo Ebola virus and the Covid-19 virus. Scientists and doctors work together to identify how the disease is transmitted so that they can then control it and prevent it from spreading. Scientists also work together to develop cures so they can treat those who catch the disease, and prevent others from getting it. By working in this way, scientists help save the lives of many people.

1

With a partner, research the Zika or Covid-19 virus.

a Find answers to all the questions in the question wheel.

b Share what you learn with another pair.

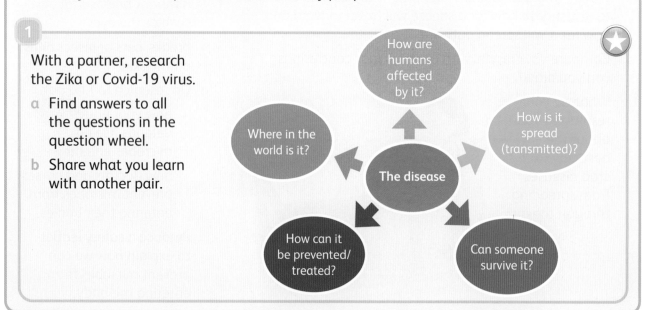

How are humans affected by it?

Where in the world is it?

How is it spread (transmitted)?

The disease

How can it be prevented/ treated?

Can someone survive it?

What have you learnt about systems and diseases?

1

a Share what you know about systems and diseases with learners in another class.

Use one of these formats: pop-up book, slideshow or an infographic (showing information in charts or diagrams).

b Include the following:
- an introduction
- information about:
  - the circulatory system
  - the respiratory system
  - the causes of infectious diseases
  - how to control the spread of diseases
  - the human defence mechanisms
  - the best way to stop diseases from spreading through the air.

If you decide to create an infographic, first research the term 'infographic' to collect ideas.

2

a Write one question you have about body systems and diseases.

b Do research or investigate to find out the answer to your question.

c Share what you find out with the class.

What can you do?

You have learnt about systems and diseases. You can:

✔ describe the function of the human circulatory system.

✔ name other vertebrates that have similar circulatory and respiratory systems to humans.

✔ describe how the circulatory system works.

✔ describe how the respiratory system works.

✔ describe how some diseases can be caused and transmitted.

✔ describe how to stop the spread of diseases through good hygiene.

✔ link ideas about the different ways that diseases can spread to good hygiene standards and human defence mechanisms that prevent microbes causing diseases.

Living things and life processes

What do you remember about life processes?

All living things, including all plants and animals, have the same seven life processes in common. Do you remember what these life processes are?

1

a. Sort the objects below into two groups: living and non-living.

b. Compare your groups with a partner's groups. Are they the same or different? Discuss the differences. Try to persuade each other where they should go.

worm   fire   cactus

whale   cloud   fig tree

human   car   bamboo

spider   Sun   Japanese maple

Let's talk

a. With a partner, list the seven life processes that all living things have in common. The first one has been done for you.

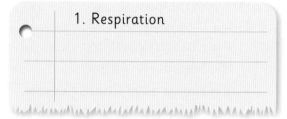

1. Respiration

b. Write everything you remember about each life process.

2

a. Write a definition for each of the seven life processes of living things.

b. Compare your definitions with a partner. What are the similarities and differences? Do you want to add anything further to your definitions?

c. Check your definitions, using the *Scientific dictionary* at the back of this book.

3

In a group, write a song that includes all seven life processes and ideas about what they involve. Sing your song to the class.

Reproduction

Think like a scientist!

Animals need a male and a female to reproduce. Some animals grow their young inside their bodies and give birth to live offspring. Other animals lay eggs.

Humans also need a male and a female to reproduce. After **puberty**, the male produces **sperm** and the female produces **eggs**. The sperm must join with the egg to create a new offspring (baby). This process is called **sexual reproduction**.

The female reproductive system contains tubes, called **Fallopian tubes**. The sperm from the male fertilises the egg from the female in the Fallopian tubes. Because this happens inside the female's body, it is called **internal fertilisation**.

Once the egg has been fertilised by the sperm, the woman is said to be pregnant. The offspring (baby) grows inside the mother's **uterus** for the **gestation period**, which for humans is about 40 weeks. The offspring (child) **inherits** (obtain) characteristics from both parents.

1

You will need...
- timer or stopwatch

a   With a partner, study these images of the male and female reproductive systems. When you are ready, cover the images.

b   Take turns. Time your partner. See how quickly your partner can name all the different parts of the human reproductive system. Write down the parts they forget.

c   Have several turns each. See how quickly you can each name every part of the human reproductive system.

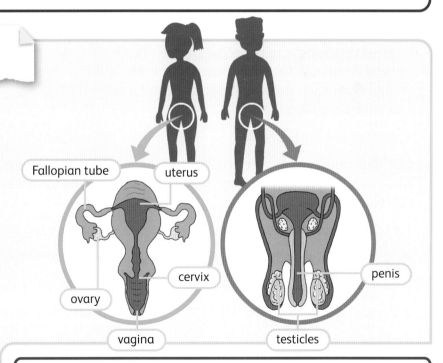

Fallopian tube | uterus | cervix | ovary | vagina | penis | testicles

Scientific words

puberty	sperm	eggs
sexual reproduction	Fallopian tubes	internal fertilisation
uterus	gestation period	inherits

Puberty

Think like a scientist!

During adolescence, children grow, and their bodies develop and change as they become adults. This stage of life is called puberty. Puberty is the time when humans reach **sexual maturity**. This means that their bodies become capable of reproduction. Some changes happen internally (inside) and others externally (on the outside).

A major change in females during puberty is that **menstruation** (periods) starts. For most women, menstruation ends around the age of 55. Usually, girls begin puberty around 10–11 years, and end puberty around 15–17 years. Usually, boys begin puberty around 11–12 years, and end puberty around 16–17 years.

Scientific words
sexual maturity
menstruation

1

a   As a group, find out about the physical changes that happen in the male and female body during puberty.

b   On a large sheet of paper, draw a two-circle Venn diagram. Label one circle male and the other female.

c   Discuss each statement below. Decide if each change happens to males, females, or both. Then write each statement into the correct section of your Venn diagram.

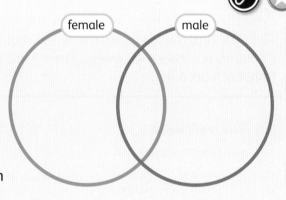

increased levels of sweat	increased levels of body odour
hair starts to grow on different parts of the body including arms, legs, face, and external reproductive organs	Adam's apple (lump in the throat) may grow and be more visible
mood changes may happen	body can go through large growth spurts
menstruation (periods) will start	acne (spots) may appear on the skin
voice changes and gets deeper	breasts develop and grow
penis and testicles get bigger	chest and shoulders get broader
hips get wider	sperm is produced

Science in context

More grandparents than grandchildren

For the first time ever, scientists say there are more elderly people than young children in the world. There are more than 705 million people over 65 years old, and about 680 million children between 0–4 years old. Scientists predict that by 2050 there will be more than two persons aged over 65 years for every person aged 0–4 years.

Why is this a problem? As a **species**, we must produce enough offspring to replace those people who die. This is called the **replacement level**. If we do not meet the replacement level, then the world population will start to shrink. Some countries have strict limits on how many children a family is allowed to have. Scientists say that for each country to achieve its replacement level, the average **fertility rate** (the number of children a woman gives birth to in her lifetime) must be about 2.1 children.

Although fertility rates are decreasing across the world, the world population is still growing. This is because the average **life expectancy** (the length of time a person lives) is increasing and death rates are decreasing. More people are living longer.

1

You will need...
- atlas

a Look at this map. It shows life expectancy at birth, from 2015–2020.

b Use an atlas to answer these questions:
- Which continent has the lowest life expectancy?
- What is the life expectancy in Australia?
- What three reasons can explain why people live longer in Canada than they do in Nigeria?

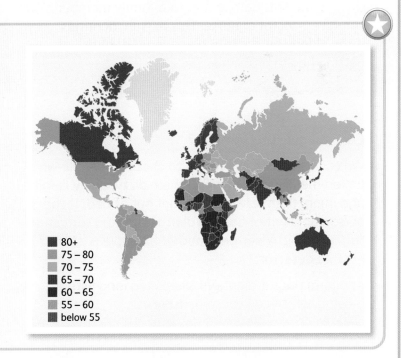

- 80+
- 75 – 80
- 70 – 75
- 65 – 70
- 60 – 65
- 55 – 60
- below 55

Scientific words

species   replacement level   fertility rate   life expectancy

Surapa Thiemjarus is a scientist who has won prizes for her work. Read about what she does.

What do you do?

I help elderly people live longer and safer lives. I designed a system that links the **Internet of Things** technologies with **Body Sensor Networks**. The *Internet of Things* includes the different devices and appliances we use every day, that are linked to the internet, such as cell phones.

Body Sensor Networks includes sensors (measuring devices) that we attach to a person. I place the sensors on people and then link the sensors to the *Internet of Things*. The sensors collect data about the person. This data is sent to a family member, caregiver, doctor or nurse. This person identifies if the elderly person is in any danger or at risk of getting health problems.

Why is this work important?

The tiny sensors help us to check on the elderly person all the time, and raise an alarm if anything is wrong. This could be the first time in the world where *Body Sensor Networks* and *Internet of Things* technologies is used to prevent harm, and to increase how quickly we can help a person.

Can you give us examples?

We can use a bed exit alarm to check on people who are on medicines that make them dizzy, and will fall and injure themselves if they get out of bed. If the person gets out of bed, an alarm rings and someone can go help the person. Another example is a sleep posture detector.

An alarm rings if a patient sleeps too long without moving, which can cause bedsores.

What are the next steps in your research?

We have the system. Now we must decide on the best, simplest and cheapest way to use it and make it available to people in their homes or in nursing homes.

Scientific words

Internet of Things   Body Sensor Network

What have you learnt about human reproduction?

1

a   Work in a small group to create a game called, *Three piles*.

b   First, make key word cards. Go back through this unit. Write each *Scientific word* and other terms from the unit on a separate piece of paper or card.

c   Then make question cards. On different cards or pieces of paper, write a question that can be answered using one of your key word cards. Challenge yourself to use other key words in your question.

d   Make sure you have a question card to match every key word card. Make sure that only one key word card can be used to answer each question.

e   Write labels for three piles: *Yes, No, Unsure*.

f   Two groups play the game. Stack all your questions together, and lay out all your key word cards in front of the other group.

g   Each group has a chance to turn over one of the other group's question cards and read it out to their group. If the group gives the correct answer, they place the question and key word cards on the *Yes* pile. If they do not know the answer, they place the question card on the *No* pile. If they are not sure, they place the question card on the *Unsure* pile.

h   Once all question cards are turned over and placed on the three piles, work together to move questions from the *No* and *Unsure* piles to the *Yes* pile. Discuss any remaining questions on the *Unsure* pile with the class.

2

Draw three pictures to summarise the unit. Write what they are and why they are important to you. Share your drawings with others.

What can you do?

You have learnt about human reproduction. You can:

✔   name the parts of the male and female reproductive systems.

✔   describe the different life stages that humans go through.

✔   describe the physical changes that take place during puberty in males and females.

✔   explain why it is important for all animals to reproduce.

Food chains

What do you remember about food chains and habitats?

A **food chain** is a way to show how **organisms** in a **habitat** feed on one another.
It uses pictures and/or words to represent the organisms, with arrows linking them.
Use the pictures below to draw your own African **savanna** food chains, with arrows.

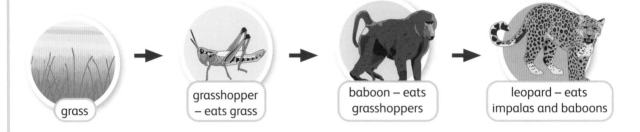

grass → grasshopper – eats grass → baboon – eats grasshoppers → leopard – eats impalas and baboons

1

a With a partner, create a mind map that includes all the words in the *Scientific words* box. Challenge yourselves to also include these words:

(producers) (energy) (prey)

(consumer) (predators)

b Compare your mind map with another group's. Can you add more ideas to yours?

Scientific words
food chain
organisms
habitat
savanna

2

You will need...
- cups you can draw and write on
- pictures of animals
- tape
- pen

a Create one cup for each of the different organisms in your food chain.

b Write as much information as you can about each organism on the cup. Use scientific words.

c Explain how your food chain works. Show this by stacking the cups.

Each time you stack a cup to show what has been consumed, make a chomping sound to indicate it has been eaten!

Food webs

Think like a scientist!

An **ecosystem** is an area made up of living and non-living things that relate to or **interact** with each other. Many organisms (animals) in an ecosystem eat more than one type of food. A **food web** is made up of several different food chains linked together. A food web shows how the living things in a habitat rely on one another for food.

The food web below shows some organisms that live in a pond. The arrows create a food web that shows the **feeding relationships** between the organisms. The arrows show who has eaten what, with each arrow pointing from the plant or animal that has been eaten to the organism that consumed it.

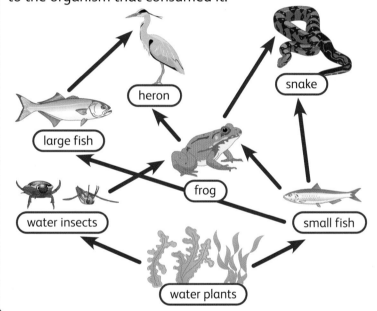

1

This food chain was made using the food web in the *Think like a scientist!* box above. Create four more food chains, using the same food web.

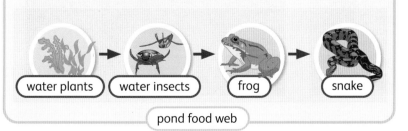

pond food web

Let's talk

a Which organisms in the pond food web are the producers, consumers or predators?

b For each predator, identify its prey.

Scientific words

ecosystem   interact
food web
feeding relationships

2

a Choose some organisms found in a garden habitat. As a class, build up the garden food chains into a food web.

b Your teacher will explain how you are going to use string to show the feeding relationships as you act out being the producers, consumers and predators.

c See how many different food chains you can make in your ecosystem.

d Discuss these questions:
- How do the links between the different food chains make a food web?
- What does the string in the food web model represent?
- Is this a good model for showing a food web? Explain.

Energy in ecosystems

Think like a scientist!

Within a habitat, there are food webs made up of many different food chains. These complex feeding relationships show how organisms (plants and animals) can have different sources of food, and how they are dependent on each other.

Plants and animals are the two types of living things found in an ecosystem. Non-living things, such as water, the weather, soil, Sun and climate, can also change and affect ecosystems. The main source of **energy** in almost all ecosystems is the Sun.

Plants capture the energy from the Sun. The energy then moves from one organism to another through the ecosystem, as each organism consumes food to keep it alive. The arrows in a food chain and food web are important because they show us where the energy moves.

Did you know?

Most of the energy consumed by an organism is lost as heat. This is the main reason why there are only a few food chains that have more than five levels of consumers.

Scientific word

energy

Let's talk

a Work with a partner or group. List some of the living and non-living things that exist in each ecosystem in these pictures.

b Choose one of the ecosystems in these pictures. Discuss what might happen if one of the non-living characteristics you identified in question **a** changed.

river

town

forest

Changes in ecosystems

Think like a scientist!

Ecosystems can change over time. Scientists gather evidence about different living organisms within ecosystems and watch how the numbers change over time.

1

a Work as a group of scientists.
Look at the information below about the feeding habits of woodland animals.

- *Hedgehogs* eat *slugs* and *beetles*.
- *Ladybirds* eat *greenfly*.
- *Thrushes* eat *snails* and some types of *caterpillar*.
- Some types of *beetle* feed on sap from plants, while others eat *greenfly* and *ants*.
- *Foxes* eat *thrushes* and *slugs*.
- *Snails* and *slugs* eat most types of leaves.
- *Greenfly* feed on sap from plants.
- *Ants* eat all types of plant parts.
- Trees, flowers and grasses trap the Sun's energy to make food.

| hedgehog | ladybird | thrush | fox |

| beetle | snail | slug | greenfly |

b Another name for a food web is an **energy flow diagram** because it shows where the energy moves to as the animals feed. Construct a food web for the woodland ecosystem above, showing where the energy moves to.

c Compare your diagram with another group's. How are your diagrams the same or different?

2

Scientists collected some data for the woodland ecosystem from Activity 1. For large animals, they counted how many there were in total. For smaller animals, they counted how many there were in a single square metre (m²).

The scientists were unable to gather data for ladybirds, beetles and greenfly. Copy the table, and add rows for ladybirds, beetles and greenfly. Decide on a value for ladybirds, beetles and greenfly.
Give a reason for your choice.

Animal	Number
foxes	2
hedgehogs	4
slugs	11 per m^2
snails	16 per m^2
thrushes	3

Scientific word
energy flow diagram

3

A new housing estate was built on the hill above the woodland. This changed the ecosystem. Scientists counted the animals at the same time each year, using the same areas each time. After three years, they recorded a decrease in the number of some woodland animals and an increase for others.

One theory was that a chemical that kills snails and slugs had got into the soil water from the housing estate gardens, and had leaked downhill into the woodland.

If this were the reason for the change in the animal numbers, what numbers would you expect of each animal? Why?

Toxic materials in ecosystems

Think like a scientist!

There are lots of different substances within any ecosystem. Many substances are helpful, but some are harmful.

Pollution happens when there are harmful substances in an environment that can damage it. Some pollution occurs naturally, such as ash in the air from a volcano. Other pollution is created by humans, such as waste gases from car exhaust pipes and the chimneys of coal-burning power plants. Some factories also release harmful waste products into the air and water from processes they use to make different goods.

Harmful substances in an ecosystem can be **toxic** (poisonous) to the organisms that live in it. Some toxic substances can **decay**, and become less harmful to the ecosystem. Others do not decay and can remain in the ecosystem.

New substances can also be introduced into an ecosystem in many ways. For example, farmers may spray crops with chemicals to help them grow strong and healthy, and these chemicals are then washed into rivers by the rain.

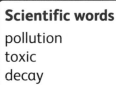

Humans pollute ecosystems in many ways that harm the plants and animals living there.

Scientific words

pollution
toxic
decay

Let's talk

Look at the picture below. Discuss how many different ways you think harmful substances are being introduced into this ecosystem.

Changes in ecosystems

1

Match the pictures with an example of how humans might add a toxic substance into an ecosystem.

A

B

C

D

E

F

G

H

I

J

rain animal waste motorcycle oil leak washing a car on a road

someone applying suncream fertilisers and pesticides broken sewer pipe

disposing of unwanted chemicals household cleaning plant waste from gardening

2

a For each example in Activity 1, think of one way humans can reduce the number of toxic substances entering the ecosystem.

b Choose one example from Activity 1. Produce an information leaflet to tell visitors to your school how they can help. Display your posters at the entrance to your school.

Challenge yourself!

Do research into the problems humans have caused by using two toxic substances – mercury and DDT (an insecticide).

- Which organisms have been damaged by these toxic substances?
- What problems have these toxic substances caused?

Science in context

More plastic than fish!

Andreas Fath, a German chemist, swam the entire length of the Tennessee River in the USA – 1 049 kilometres (652 miles) – in just 34 days! Why did he choose to do this enormous task? Because he wanted to raise awareness about water pollution (and get his name in the *Guinness Book of World Records*).

water pollution

Fath worked with the environmental scientist Martin Knoll to organise the swim. Both scientists are worried about the health of rivers, oceans and other bodies of water because of the amount of plastic materials found in them. In 2016, a study was done which said that within 35 years the oceans could hold more plastic (by mass) than fish. Scientists argue about whether this prediction is true, but they all agree that plastic pollution of water is an issue we must tackle. Pollution harms our oceans and rivers, as well as all the living things that depend on these bodies of water.

Plastic is used in many products, from packaging to plastic bags. Massive amounts of plastic end up in the ocean. This plastic contains harmful substances that can be toxic to wildlife that live in the ocean. Animals can also get tangled in the plastic waste, which can harm or kill them.

Let's talk

Discuss these questions with a partner:

a Why is it important to have clean water?

b Have you seen water that you thought was **polluted**? Where?

c What evidence was there that the water was polluted?

d Where do you think the **pollutants** might have come from?

2

a As a group, write what you will do to clean the 'polluted' water you made in Activity 1.

b Make a list of the equipment you will need.

c How will you know if your method has worked?

d Carry out the test you planned.

e Share your results with another group.

1

You will need...
- cup
- salt
- sugar
- spoon
- drinking water

a Pour some drinking water into a cup.

b Add 'pollutants' to the water – half a teaspoon of salt and half a teaspoon of sugar. Stir.

c Discuss how you could remove the 'pollutants' from the water.

Scientific words

polluted   pollutants

Toxic accumulation in food chains

Think like a scientist!

If a plant or an animal in an ecosystem absorbs a toxic substance, the toxic substance can enter the food chain. If the toxic substance cannot be broken down or **excreted** by the organism, the toxic substance remains inside it and can damage the organism.

If another organism eats that organism, the toxic substance then moves through the food chain. All organisms in the food chain or food web are then at risk of being damaged.

The build-up of toxic substances in a food chain or web is called toxic **accumulation**. This build-up happens over time, and the organisms who are most likely to be damaged are the predators, who are at the end of the chain.

Scientific words

excreted
accumulation

1

You will need...

- a cup with the name of an organism written on it
- red, green, yellow counters (represent food)
- 'feeding' bags with the counters, placed around the 'ocean'
- waste box

a Play the *Ocean game*. The aim is to gather food to survive. Stand around the edge of the 'ocean' (classroom). Your teacher will give you a cup with the name of the animal you are, in the ocean ecosystem. Your teacher will explain the instructions to you.

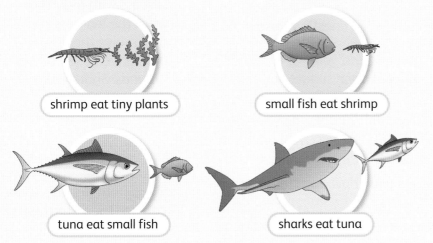

shrimp eat tiny plants

small fish eat shrimp

tuna eat small fish

sharks eat tuna

Let's talk

Discuss with a partner how the *Ocean game* explains what happens to toxic materials in the ocean ecosystem. Where do toxic materials accumulate in the ocean ecosystem?

Work safely!

Do not play this game if social distancing is in force. Make sure you only walk, and move safely around the room. Keep your eyes open to avoid bumping into others. You only need to touch someone gently.

Scientists protect ecosystems

Think like a scientist!

Some human activities result in harmful substances entering ecosystems. Many people around the world are working to protect ecosystems. Scientists who work to **conserve** (protect) ecosystems are called **conservationists**.

a   What sorts of things do conservationists do?

b   Why do they do these things?

Scientist searching for new rainforest species

What can you do to help to protect ecosystems?

Scientific words

conserve
conservationists

Let's talk

Some Class 6 learners are concerned about litter on their school grounds. They know that litter can add toxic substances, and is a danger to wildlife. They know that litter attracts pests that can spread diseases.

Discuss with a partner what the learners could do to solve the litter problem on their school grounds. Share your ideas with the class.

1

You will need…
- map of your school grounds

a   Work with a partner. List ideas about what you can do to improve your school grounds as an ecosystem for living things. Here are some suggestions:

litter bins   wildflower area   'bug hotel'

composting area   vegetable garden

pond   trees   bird feeders   nesting boxes

b   Use the map of your school grounds and your list of ideas to produce a dot plot of the number of each suggestion that already exists.

c   Using your map of the school grounds and the data in your dot plot, draw a new map showing how the school grounds could be improved as an ecosystem for living things.

d   Add labels to your map which explain how each feature supports living things.

e   Draw a food web to show the possible impact of the changes.

2

As a class, decide which ideas from Activity 1 you can do. Then put them into action.

What have you learnt about ecosystems?

1

Use what you found out about in this unit to encourage learners in other classes to care for ecosystems.

You could give a presentation, create a poster or leaflet, film a video, design and make a game, or use other ideas.

2

a As a group, create a mind map. Write *Ecosystems* in the middle of a large sheet of paper.

b Each learner uses a different coloured pen to add on the mind map what they have learnt in this unit.

c Tick (✓) if you agree with someone else's point. Add a question mark (?) if you are unsure. Put a cross (✗) if you disagree. Make as many links as you can with each others' ideas.

d What is the most interesting thing you learnt about ecosystems? Explain why.

3

What else would you like to find out about caring for ecosystems?

a Write one question to which you would like to know the answer.

b Do research to find out the answer to your question.

c Share what you have found out with the class.

What can you do?

You have learnt about ecosystems. You can:

✔ identify several different food chains in one ecosystem.

✔ explain how a food web is different to a food chain.

✔ state where all animals on Earth get their energy from.

✔ explain how energy is passed through a food chain, and how to represent this in a diagram.

✔ describe ways in which human activities add toxic substances to an ecosystem.

✔ explain why toxic substances are more harmful to predators at the end of a food chain.

Reversible and irreversible changes

Materials

What do you remember about materials and how they change?

Matter makes up everything around us. Scientists define matter as anything that has volume (takes up space) and mass. Matter exists in one of three states: solid, liquid or gas. When scientists talk about **materials**, they mean the types of matter from which objects are made. Examples are the wood in a tree, the aluminium in a can, the oxygen in bubbles.

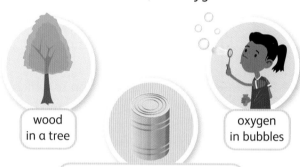

wood in a tree

oxygen in bubbles

aluminium used to make a can

When materials are heated or cooled enough, they may change state:

• If a solid is heated enough, it may change into a liquid.

• If a liquid is cooled enough, it may change into a solid.

• If a liquid is heated enough, it may change into a gas.

• If a gas is cooled enough, it may change into a liquid.

Let's talk

Talk about how you would define each word in the *Scientific words* box. Discuss your ideas with a partner.

1

Write which state or **states of matter** have each of the properties below. Discuss your ideas with a partner. There are two examples for you.

a can be poured – liquids and gases
b keep their shape – solids
c fill up spaces
d take the shape of their container
e usually invisible
f can be cut or shaped
g move around

2

Name the material(s) each object below is made from. Say whether it is a solid, liquid or gas. There is one example for you.

A balloon – rubber (solid), air (gas)

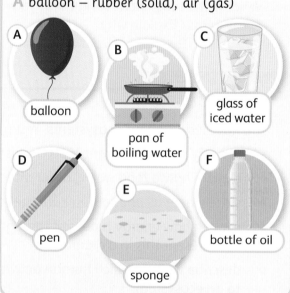

A balloon

B pan of boiling water

C glass of iced water

D pen

E sponge

F bottle of oil

Scientific words

matter   materials   states of matter

Changing materials

1

Say what is happening in each picture below. Use these words:

condensation   evaporation   freezing   melting

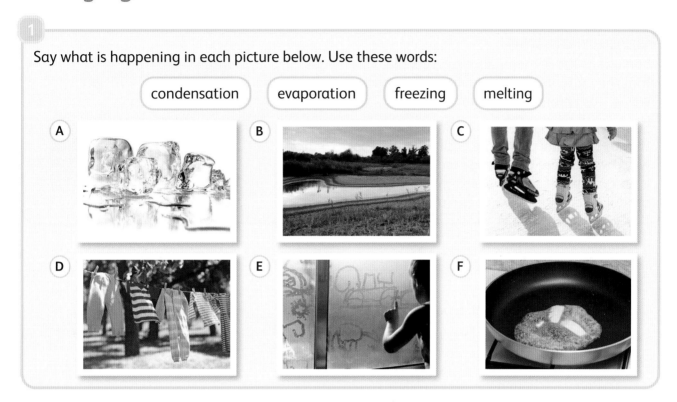

2

Name the processes labelled **a** to **d** in the diagram below. Use these words:

condensation   evaporation

freezing   melting

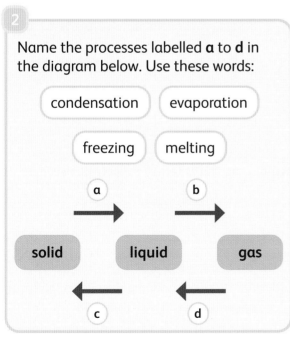

Scientific words

condensation   evaporation
freezing   melting

3

An acrostic is a composition in which certain letters in each line form a word or words, written downwards. Here is an example, using the word Newton.

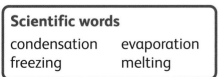

N oteworthy mathematician

E xplains gravity,

W hite light, and

T he laws

O f the

N atural world.

Use one of the words in the *Scientific words* box to make an acrostic showing what you know about materials and how they change.

Reversible or irreversible?

Think like a scientist!

Some changes in materials are **reversible**. This means that the material can be easily changed back to the way it was before the change took place.

Some changes in materials are **irreversible**. This means that the material cannot be easily changed back to the way it was before the change took place. Baking bread is an example of an irreversible change. A loaf of bread cannot be changed back to raw dough.

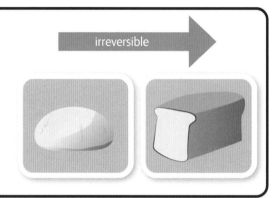

irreversible

Let's talk

Discuss these questions with a partner:

a What change is taking place in each set of pictures? Use these words:

baking   burning   dissolving

freezing   melting   rusting

b Is each change reversible or irreversible? If the change is reversible, what could you do to reverse it? Share your ideas.

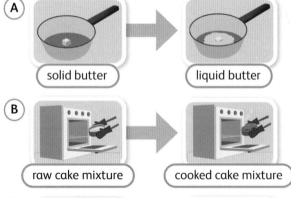

(A) solid butter → liquid butter

(B) raw cake mixture → cooked cake mixture

(C) liquid lolly → solid lolly

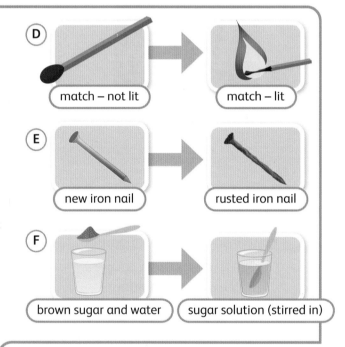

(D) match – not lit → match – lit

(E) new iron nail → rusted iron nail

(F) brown sugar and water → sugar solution (stirred in)

1

Draw an example of each change below. Label the change *reversible* or *irreversible*.

a condensation   b toasting

c rotting   d evaporation

Scientific words

reversible   irreversible

Reversible changes

Think like a scientist!

A reversible change is a change that can be easily **reversed** (undone or changed back). Sometimes scientists call reversible changes, **physical changes**. A physical change alters the **physical properties** of a material (how it looks or feels), but it does not produce any new materials.

Because no new materials are produced, the material can be changed back to how it was before. Reversible changes include dissolving, freezing, melting, **vaporisation** (evaporation or **boiling**) and condensation.

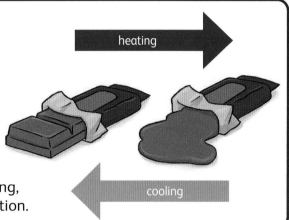

heating

cooling

1

Copy and complete the table below about reversible changes. There is one example for you.

Reversible change	It happens when ...	It can be reversed by ...
a freezing	a liquid is cooled and becomes a solid	heating the solid so that it turns back into a liquid
b melting		
c vaporisation		
d condensation		

2

a   Work in a group. Choose an example of a reversible change.

b   Collect the materials and equipment you will need.

c   Demonstrate the reversible change to another group. Demonstrate changing the material back to the way it was before the change took place. You are responsible for the way you should work.

Let's talk

Observe the material you changed in Activity 2. Discuss these questions in your group:

a   Is the material exactly the same now as it was before you did the activity? If not, how is it different?

b   If there are differences, is the change you demonstrated really reversible? Explain your thinking.

Scientific words

reversed   physical changes
physical properties   vaporisation   boiling

Melting and freezing

Think like a scientist!

Melting and freezing are reversible changes. Melting is when a material changes state from a solid to a liquid. You can make a solid melt by heating it. The **melting point** of a solid is the temperature at which it melts. This is different for every material.

Freezing is the reverse of melting. Freezing is when a material changes state from a liquid to a solid. You can make a liquid freeze by cooling it. The **freezing point** of a liquid is the temperature at which it freezes. This is different for every material.

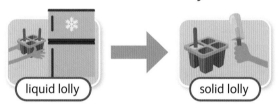

liquid lolly   solid lolly

All materials **conduct** heat (allow heat to pass through them) at different rates. This is called **thermal conductivity**.

1

Study this table and then answer the questions below.

Material	Freezing point (°C)
petrol	−50
mercury	−39
linseed oil	−20
castor oil	−10
water	0
olive oil	3
palm oil	24
coconut oil	25

a Room temperature is about 22 °C. Name three materials that are liquids at room temperature, and two materials that are solids.

b The temperature in a household freezer is −18 °C. Name three materials that are liquids at room temperature, but will freeze in a household freezer. Name three materials that will remain liquid in a household freezer.

2

Jed sells ice, but it melts before he can sell it. In a group, discuss these questions and write your answers:

a How can Jed make sure that the temperature in his freezer box does not go above melting point?

b How could you test your ideas? What do you predict?

c What measurements will you make?

d What equipment will you need?

e What **variables** must you take into account to make sure the test is fair?

3

a Carry out the test you planned in Activity 2.

b What pattern can you see in your results?

c Do your results support your prediction? Explain your results scientifically.

d What conclusion can you draw?

e Make another prediction based on your results.

Scientific words

melting point   freezing point   conduct
thermal conductivity   variables

More about melting

1

Some Class 6 learners carried out a test to discover how the temperature of ice changes as it is heated. They took crushed ice out of the freezer. They used a digital thermometer to measure the temperature of the ice, 1 minute apart. Answer these questions about the **line graph** of their results:

a   Describe the pattern of the line graph over time.

b   Describe what is happening to the ice in each section of the line graph.

c   Predict the temperature after 15 minutes. Explain your answer.

d   What was happening to the ice between 4 to 6 minutes? Use data from the graph to explain.

e   What was happening to the ice after 10 minutes? Use data from the graph to explain.

f   One learner argued that it is possible to heat a substance without the temperature rising. Do you agree or disagree? What are your reasons for your answer?

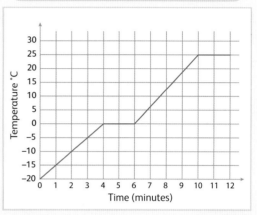

Line graph to show how the temperature of ice changes as it is heated

How does the pattern of the line graph link to the melting point of ice?

Let's talk

Choose one material for each of these properties (for example, margarine, sesame oil, ghee):

• one that is solid when kept in the fridge
• one that is liquid at room temperature
• one that in its solid form is soft enough to put a thermometer inside.

Discuss these questions in a group:

a   Are the melting points of your chosen materials the same or different? What do you predict? Explain your prediction using your scientific knowledge about materials.

b   What could you do to test your ideas?

c   What type of measurements will you need to make?

d   Will you need to repeat your measurements? Explain your answer.

2

a   Write a plan of the test you discussed in the *Let's talk* activity. Include a labelled diagram showing what you will do.

b   Carry out the test you planned.

c   Present your results using a line graph.

d   Explain whether your results support your prediction.

e   What patterns are in your results?

f   Are there any results that do not fit the pattern? What do you think caused this?

Scientific word

line graph

47

Evaporation, boiling and condensation

Let's talk

Discuss what you remember about evaporation and condensation.

a Where has the dew come from?

b Dew forming on grass is an example of which process?

c The dew will evaporate to become which gas?

d What factors will affect the speed at which the dew evaporates?

dew on grass

1

Use the *Think like a scientist!* box to help you answer these questions.

a Where does evaporation take place?

b At what temperature does evaporation take place?

c What are the similarities and differences between evaporation and boiling?

d What happens during the process of condensation?

e What has to happen to a gas before it condenses?

Scientific word

boil

Think like a scientist!

Evaporation and condensation are reversible changes. Evaporation is when a material changes state from a liquid to a gas. It takes place on the surface of a liquid.

Evaporation takes place at different temperatures for different materials. You can make a liquid evaporate more quickly by heating it. If you heat a liquid to a high enough temperature it will **boil**.

When liquid boils, evaporation takes place throughout the liquid, not just on its surface. You can see bubbles of gas in a pan of boiling water.

evaporation

condensation

Condensation is the reverse of evaporation and boiling. Condensation is when a material changes state from a gas to a liquid. A gas condenses if it is cooled enough. A gas may condense when it comes into contact with a cold surface.

2

a Work in a group. On a large sheet of paper, write two headings: *Evaporation* and *Condensation*. Write as many examples of each process as you can.

b Swap sheets with another group. Comment on the other group's work. Say something that you thought was good and something they could find out more about.

Evaporation

Think like a scientist!

Evaporation happens when a liquid slowly changes state to become a gas. When water evaporates, it becomes a gas called **water vapour**.

1 You will need...
- measuring jug
- saucer
- water

Investigate what happens to water when you leave it outside. Start the investigation to observe water evaporation in the morning so that you can complete it the same day.

a Pour 100 ml water into the saucer.

b Place the saucer in a warm place – outside in direct sunlight or on a sunny windowsill.

c A few hours later, measure the volume of water in the saucer. What do you observe?

d Record your observations.

e Explain your observations, using the correct vocabulary.

Let's talk

With a partner, discuss which pictures show water evaporating to become water vapour.

 A

 B

 C

 D

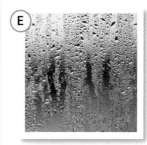 E

 F

Let's talk

Discuss these questions with a partner:

a Predict: In the above *Let's talk* activity, will the water evaporate faster in a sunny place or in a shady place? Why?

b What test could you do to find out the answer?

c How could you make sure that the test is fair?

Scientific word
water vapour

2
a Carry out the test you discussed in the *Let's talk* activity.

b Describe the results of the test.

c Was your prediction correct? Use your scientific knowledge to explain why or why not.

d What type of scientific enquiry is Activity 1 and 2? Discuss what features of the enquiries make you think this.

Evaporation, airspeed and temperature

1

You will need...
- electronic scale
- two washcloths
- bowl of water
- washing line
- clothes pegs
- electric fan
- circuit breaker
- mains electric socket

Does washing dry more quickly on windy days than it does on still days?

a Discuss how the speed or movement of the air (wind) affects how quickly water evaporates.

b Write your predictions.

c Measure the mass of each dry washcloth. Record each mass.

d Soak both washcloths for the same length of time in the bowl of water. Wring out the water from both washcloths.

e Measure the mass of each wrung-out washcloth. Record its mass.

Work safely! ⚠️

For safety, the washing line should be above your head, but low enough to reach without stretching. Use an electric fan with a circuit breaker. Only use electric sockets with dry hands.

f Fix the washing line somewhere safe in the classroom. Peg the washcloths on the line.

g Place the electric fan so that it faces one of the washcloths. Plug it into the mains electric socket and switch it on.

h Measure the mass of each washcloth every 30 minutes. Record each mass.

i Do the washcloths dry at different speeds? If so, which washcloth dried faster? Explain why.

2

a In a group, discuss how the temperature of the air affects how quickly water evaporates.

b Write your predictions.

c Plan and carry out a fair test.

d Decide how you will present your results.

Let's talk

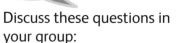

Discuss these questions in your group:

a What have you found out from doing the tests? Were your predictions correct?

b How accurate were your tests?

c If you did your tests again, what would you do differently? What would you improve? How would this help you collect more accurate results?

d Do you think you collected enough data to draw a conclusion?

e Rank these days into the ones you think will dry washing the best. Explain your reasons.
- windy and cold days
- windy and sunny days
- still and cold days
- still and sunny days

f What type of scientific enquiry is Activity 1? What features of the enquiry make you think this?

Evaporation and surface area

Let's talk

Each container holds the same volume of water. The **surface area** (size of the top) of the water is different in each container.

Discuss these questions with a partner:

a Which container has the largest surface area?

b Which container has the smallest surface area?

c From which container do you think water will evaporate fastest? Why?

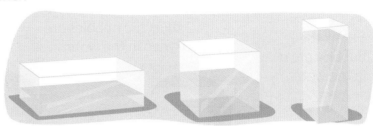

1

Did you know?

Area = length x width

Area is measured in square units (mm^2 or cm^2).

You will need...
- different sized rectangular containers with different surface areas
- sticky labels
- pencil
- ruler
- measuring jug
- water

Investigate how surface area affects the speed of evaporation.

a Label each container with numbers or letters for easy identification.

b Pour the same volume of water into each container.

c Calculate and record the surface area of the water in each container. Read the *Did you know?* box.

d Leave the containers together in a sunny place.

e After a few days, measure the volume of water that is left in each container.

f What conclusion can you draw from your results?

Think about what evaporates from your body.

Challenge yourself!

If you lick your finger and move it in the air, it feels colder. Why do you think this happens?

Scientific word

surface area

Boiling

Some Class 6 learners are discussing the bubbles they can see in a kettle of boiling water. What do you think of their ideas? Discuss your ideas with a group.

I think the bubbles are air in the water.

I think the bubbles are oxygen in the water.

I think the bubbles are water that has turned into a gas.

I think the bubbles are water vapour in the water.

I am not sure what the bubbles are.

Anne

Sabrina

Alan

John

Jasmine

Let's talk

Look at the picture of the boiling kettle. Discuss these questions:

a What must you do to water to make it boil?

b Where does boiling occur in a liquid?

c Which gas forms bubbles in boiling water?

d Why is there a gap (space) between the kettle spout and the condensed cloud of water droplets?

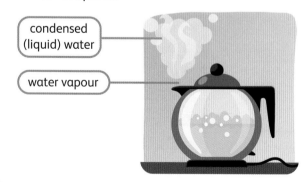

condensed (liquid) water

water vapour

Think like a scientist!

Water boils when we heat it to a high enough temperature. This happens throughout the liquid, and not just on the surface. When water boils, it changes to a gas called water vapour. The bubbles in boiling water are water that has turned into a gas, so we could say they are water vapour. Water vapour is invisible. This gas can also be called steam. It is in the kettle above the surface of the water. As water vapour escapes from the kettle, it meets the cooler air outside. Some of the water vapour condenses into tiny water droplets that form a cloud. The water vapour that is escaping from the kettle takes a little time to cool down and condense. This is why there is a gap between the kettle spout and the cloud of water droplets.

Investigate boiling

Work safely!

ONLY an adult should carry out the activities on this page. Stand well back while observing.

1

The temperature at which water boils is its boiling point. What is the boiling point of water? This is a very important number in Science! Remember it!

2

Your teacher will demonstrate different activities to show how temperature, water volume and salt affect how quickly water boils. Your teacher will place two identical pans of water on two identical hot plates to boil. The heat under pan A is higher than the heat under pan B.

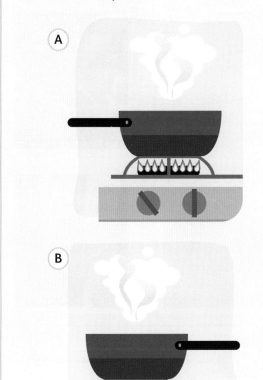

Predict what will happen to the temperature of the water in the two pans. In which pan will the water boil first?

3

You have watched your teacher carry out the fair test.

a Was your prediction in Activity 2 correct? If not, try to explain why.

b Draw a diagram to show what happened in the demonstration. Add labels and a caption.

4

a Predict: Will water in a smaller pan reach boiling point slower than in a larger pan? Why do you think this is? Your teacher will demonstrate. Time how long it takes for water to boil in different sized pans.

b Predict: Will salt water boil at the same temperature as freshwater? Your teacher will demonstrate finding the boiling point of salt water.

c Were your predictions correct?

d Ask four questions about the amount of water being boiled, and the effects of salt on the boiling point of water. Use these question starters: What if …? How does …? Which …? How much …?

Think like a scientist!

You know that the boiling point of water is 100 °C. Boiling water does not get hotter than 100 °C, even if it boils for longer. Adding salt to water makes it boil at a higher temperature.

Condensation

Think like a scientist!

Condensation happens when a gas cools. It then changes state into a liquid. When water vapour (water as a gas) condenses, it becomes liquid water.

Work safely!

Only an adult should carry out this activity. Stand well back while observing.

1

Your teacher will use an oven glove to hold a cold mirror above a saucepan of boiling water, so that you can observe water changing from a gas to a liquid.

a Observe the surface of the mirror. What happens? Why?

b Why does your teacher need an oven glove to hold the mirror?

2

Observe how the temperature of an object affects the condensation of water vapour.

a Use two identical drink cans. Leave one can in the classroom. Put the other can in the fridge for a few hours.

b Take the can out of the fridge. Put the two cans side by side and observe them for a few minutes.

c Record your observations.

Let's talk

Which pictures show water vapour condensing to become water? Discuss your ideas with a partner.

A

B

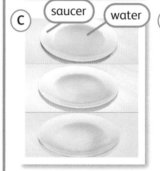
C saucer water

D

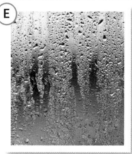
E

F

Let's talk

Discuss the cans in Activity 2 with a partner:

• What appeared on the outside of the cold can?

• Where did this come from?

• Why did this happen?

Share your ideas with another pair.

Investigate the rate of condensation

Think like a scientist!

How quickly a liquid condenses is its rate of condensation.

Work safely!

Take care when working with hot water!

Let's talk

Discuss these questions about Activity 1:

a Which surface had more water on it – was it the surface that was cooled by the ice, or the surface that was not cooled?

b Where did the water come from?

c Does cooling water vapour increase the rate of condensation? Explain your answer, based on your observations.

1

You will need...

- four clear plastic cups
- hot tap water
- ice cube
- kitchen roll
- magnifier (optional)

Investigate how the temperature of an object affects how quickly water condenses.

a With a partner, discuss how you could speed up the rate of condensation. Explain your thinking.

b Now investigate how cooling affects the rate of condensation. Fill two cups with water from the hot tap so that they are about two-thirds full.

c Place the other two cups upside down, on top of the first two cups.

d Place an ice cube on top of one pair of cups. Wait a few minutes.

e Remove the ice. Use a paper towel to dry the part of the cup where the ice was.

f Carefully observe the inside surface of the base of the two top cups. Use a magnifier.

g Compare the amount of water on the inside surface of the base of the top two cups.

2

a Record your observations from Activity 1 by drawing diagrams.

b Label your diagrams to explain what you think happened. Use these key words:

water (liquid)

temperature   hot

ice (solid)   cold

water vapour (gas)

evaporation

condensation

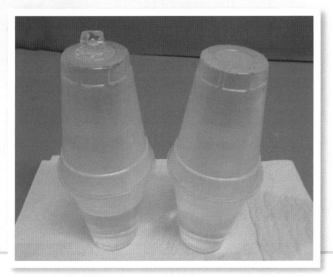

Change the rate of condensation

Think like a scientist!

You have discovered the following about water vapour and condensation:

- When water vapour in the air touches a cooler surface, it condenses to form droplets of water.
- The colder the surface that water vapour touches, the faster it condenses.

Let's talk

condensation on a mirror

Pia notices that after her shower, there is a lot of condensation on her bathroom mirror. Pia wants to find a way to slow down the rate at which condensation forms on the mirror. She has three ideas she thinks might work.

Which idea do you think will work best? Why? Discuss your ideas with a partner.

1. I could try using cooler water in the shower.
2. I could try to heat the mirror.
3. I could wet the mirror before the shower.

1

a Work in a group. Decide how you will test each idea from the *Let's talk* activity. Discuss what you will need.

b What will you keep the same to make sure the test is fair?

c What relevant observations will you make during your test?

d Discuss which idea you think will work best, and why.

e Carry out the investigation.

f Write a short report. Explain what you did and what you found out.

g Share your report with other groups.

What advice would you give Pia to reduce the amount of condensation on her bathroom mirror?

Work safely! ⚠️

An adult must be present when you heat water.

Link evaporation and condensation

Think like a scientist!

A **solar still** is a device that uses the processes of evaporation and condensation to produce clean fresh water.

1

You will need...

- cup
- water
- salt
- teaspoon
- cling film
- scissors
- large bowl
- small stone
- waterproof tape

Follow these instructions to make a solar still that will produce fresh water from salt water.

a Pour a few centimetres of water into the bowl.

b Add a few teaspoons of salt and stir well.

c Place the bowl outdoors in direct sunlight.

d Place the cup in the centre of the bowl. Take care not to splash any salt water into the cup.

e Place the cling film over the bowl. Push it down in the centre so that it curves down above the top of the cup. Make sure it does not touch the cup.

f Put waterproof tape around the edge of the cling film where it touches the side of the bowl.

g Place the small stone on top of the cling film, directly above the cup.

h Leave your solar still in sunlight for a few hours. Wait for water to collect in the cup.

i When there is water in the cup, taste it. It should not taste salty at all!

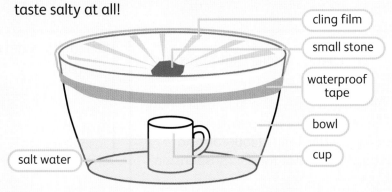

cling film
small stone
waterproof tape
bowl
cup
salt water

2

How do you think the solar still you made in Activity 1 works? Write an explanation. Use these words:

heat   the Sun

water vapour

water evaporate(d)

condense(d)

fresh water   cool

salt water **solution**

Let's talk

When might a solar still be useful? Why? Discuss your ideas with a partner. Then share them with the class.

Scientific words
solar still
solution

Dissolving

Think like a scientist!

When some solids with small particles mix with a liquid, they **dissolve**. The liquid used to dissolve a solid is called a **solvent**.

When a solid dissolves in a liquid it forms a mixture called a solution. A solution is clear. But if the solid is coloured, then it may change the colour of the solution.

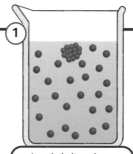

1 A solid dissolves because the particles of the solvent collide with the particles of the solid.

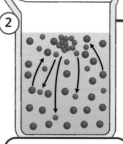

2 The solid particles gradually move away from each other until they are evenly spread through the solvent.

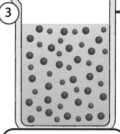

3 The solid particles are still in the solution. They are just spread out and you cannot see any particles of the solid.

1

You will need...
- large jar or container
- bag of dried lentils
- small amount of dried fruit (currants, raisins or sultanas)
- spoon

Make a model to demonstrate the idea of dissolving.

a Fill your jar with lentils by pouring them in.

b Add a spoonful of the dried fruit.

c Stir the spoonful of fruit in the lentils. What do you see happening to the fruit?

Let's talk

Think about the model you have just made in Activity 1.

a Which part is the solvent and which part is the solid that dissolves?

b What happened to the solid in the solvent?

c How does the model help to demonstrate what happens when a solid dissolves in a solvent?

d How is the model a good representation of dissolving? Explain your reasons.

e How could the model be used to explain what happens when salt is added to water, to make salt water?

Think like a scientist!

Salt forms a solution when it is mixed with water. A solid that forms a solution when it is mixed with water is **soluble**.

When a solid is mixed with a liquid but does not dissolve in it, a different kind of mixture forms. This is called a **suspension**. A suspension is cloudy. You can see particles of the solid floating in the liquid. Flour forms a suspension when it is mixed with water. A solid that forms a suspension when mixed with water is **insoluble**.

Scientific words
dissolve
solvent
soluble
suspension
insoluble

Investigate solubility

1

instant coffee coffee grounds sugar salt powder paint

laundry powder flour vitamin tablet tea cocoa

a Predict whether each material above is soluble or insoluble.

b Record your predictions with your reasons in a table.

c Collect some of the materials from the pictures. Test your predictions. Copy and complete this table, showing the material, your prediction and your observations.

Materials	Prediction	Observation:	
		Soluble	**Insoluble**
tea	insoluble		✓

d Work with a partner. Explain the reasons for your observations. Use some of these words:

dissolve solution soluble suspension insoluble

2

a Work with a partner. List the words in the *Scientific words* boxes from this unit so far.

b Invent, make and play a game to practise all these words.

c Play your game with another pair.

Challenge yourself!

Work in a group to produce role plays to explain the differences between melting and dissolving.

Temperature and solubility

1

You will need…
- three transparent plastic cups
- teaspoon
- water at three different temperatures – from a hot tap, at room temperature, and from the fridge
- measuring jug
- soluble material
- thermometer
- timer or stopwatch

Investigate how the temperature of a solvent affects how quickly a solid dissolves:
- Choose one soluble material from Activity 1 on page 59.
- Add one level teaspoon of the soluble material to water at the three different temperatures and stir it.
- Time how long it takes until your material has completely dissolved.

a First, predict which temperature water will dissolve your material the quickest. Explain your reasons. Record your predictions.

b Plan how to carry out your test and how to make sure it is fair.

c Take the temperature of the three different waters.

d Carry out your test. Record your results in a table.

e Draw a line graph of your results to show the time it took for the material to dissolve at the different temperatures.

> In the line graph, plot the temperature on the x-axis and the time on the y-axis. Label them.
>
> Think about the scale you will use on each axis.
>
> Give the graph a title.

f Was your prediction correct?

g Compare your results with a group that looked at a different soluble material. Did the different substances dissolve equally fast? How do you know?

h Use your line graph to predict how long it would take to dissolve your solid at two other temperatures that you did not test.

Did you know?

Water that is at room temperature, is water that has been left in a room long enough for the temperature of it to become the same as the temperature of the room.

Let's talk

Think about the experiment you have just carried out and what you have learnt about dissolving. Look back at page 58, at the *Think like a scientist!* boxes and the lentil model. Explain the results of your experiment. Use ideas that link the temperature of the water to ideas about the particle model.

Irreversible changes

Think like a scientist!

An irreversible change cannot be easily reversed. Scientists call irreversible changes **chemical reactions**. The materials taking part in and changing during a chemical reaction are called **reactants**.

In a chemical reaction, one or more new materials are usually formed. The new materials are called **products**. They are completely different from the original materials. Sometimes these new materials are useful.

Evidence that a chemical reaction has happened may include a gas being produced, or a change in colour or temperature.

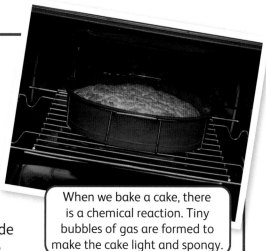

When we bake a cake, there is a chemical reaction. Tiny bubbles of gas are formed to make the cake light and spongy.

1 You will need...

- saucepan
- stove
- whole milk
- thermometer
- white vinegar
- tablespoon
- bowl
- sieve
- kitchen roll

a Pour 250 ml of whole milk into a saucepan. Heat it gently until the temperature reaches 50 °C. Use the thermometer to measure the temperature. Then remove the saucepan from the heat.

b Add one tablespoon of white vinegar and stir. Lumps of a solid will form in the mixture.

c Sieve the mixture into a bowl. Place the solid material on kitchen paper to dry.

d Squeeze the solid material together and shape it.

e Put it in a warm place for a few days to let it dry. When your material is hard, paint it.

The material you made is a polymer. Natural polymers include wool, rubber and cellulose (the main substance in wood and paper). Synthetic (human-made) polymers include plastics, such as nylon, PVC and polystyrene.

Work safely!

Do this activity under adult supervision.

Let's talk

Discuss these questions with a partner:

a Think about the change that takes place when white vinegar is added to warm milk. Is this change reversible or irreversible? How can you tell?

b Do you think it is possible to separate the mixture back into milk and vinegar? Explain.

c Identify the reactants and products.

Think about what you did in Activity 1.

At what point did the change become irreversible? Explain your thinking.

Scientific words

chemical reactions   reactants   products

Investigate temperature change

Think like a scientist!

When the reactants cement, sand, water and limestone are mixed together, a new material is produced: concrete. Concrete is the product of this chemical reaction.

You cannot turn concrete back into the materials that made it. In this reaction, there is a change in temperature.

Remember that a reversible change means that the material can be easily changed back to the way it was before the change took place.

An irreversible change means that the material cannot be easily changed back to the way it was before the change took place.

You will need...

- bicarbonate of soda
- vinegar
- thermometer
- plastic containers
- teaspoon

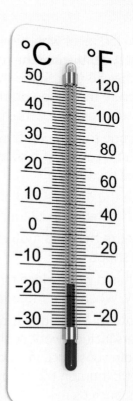

In a group, carry out an investigation where there will be a temperature change.

a Pour four tablespoons of vinegar into a container.
   Take the temperature of the vinegar.

b Mix one teaspoon of bicarbonate of soda with the vinegar.
   Take the temperature of the vinegar and bicarbonate of soda after you have mixed them together.
   Observe and note any other changes.

c Has a chemical reaction taken place? How do you know?

d Is it possible to reverse the change that has taken place? Why or why not?

e How could you speed up the change? Try out your ideas. Describe what you did and what happened.

Make bath fizzes

1

You will need...

- cupcake tray
- tablespoon, teaspoon, wooden spoon
- mixing bowl
- sieve
- cup
- pipette
- paper cake cases
- bicarbonate of soda
- citric acid
- dried flower petals such as roses (optional)
- olive oil
- essential oil
- food colouring
- water spray bottle

Make bath fizzes to observe an irreversible change.

a Rub a little olive oil onto the cupcake tray to stop the bath fizzes from sticking.

b Sieve nine tablespoons of bicarbonate of soda and three tablespoons of citric acid into the bowl. Mix together. Add a few dried flower petals if you wish.

c Pour six teaspoons of olive oil, five drops of essential oil and a few drops of food colouring into a cup (use a pipette for this).

d Slowly add the liquid from the cup into the bowl, one spoonful at a time. Mix well.

e The mixture should start to hold together, like a sandcastle. Spray on a little water if needed, but mix quickly.

f Spoon the bath fizz mixture into your cupcake tray. Press down to smooth the top.

g Put the tray in a warm place for 24 hours.

h When the bath fizzes are set, carefully turn the tray upside down. Tap out each bath fizz and put them into paper cake cases.

i Carefully add one bath fizz to warm water. What changes take place? Which of them do you think are reversible? Which are irreversible? Explain your reasons.

An irreversible change happens when you add your bath fizz to water. There is a chemical reaction between the bicarbonate of soda and the citric acid, which creates a new material – carbon dioxide. Carbon dioxide is a gas and this makes the bath fizzes fizz!

bath fizzes

Science in context

Scientists who make new materials

Materials chemists are scientists who develop materials. They use their scientific knowledge of materials and their properties to make some materials even better, or to create brand new materials. Materials chemists have created many of the items that we use today, such as soap, make-up and materials for packaging.

In 1935, a scientist called Wallace Carothers invented nylon, a strong, hardwearing material that is cheap to produce. Nylon can be made into fibres which can be woven together to make a strong, lightweight, silky fabric. Nylon is still one of the most common fabrics used today.

Hot air balloons are made from nylon.

1

Research information about materials chemists: What qualifications do they need? Where do they work? What do they do? How do they help people?

2

Find out about the work of one of these materials chemists:

Spencer Silver   Ruth Benerito

Stephanie Kwolek

a What new material did the chemist invent?

b When and how did she or he invent it?

c What properties of the material make it useful?

d What uses does the material have?

e Work with a partner who researched a different materials chemist. Take turns to role-play an interview to find out about their materials chemist.

Scientific word
materials chemists

Let's talk

Scientists have created these new materials:

• a super-sticky material

• an extremely light foam that does not conduct heat

• a type of concrete that can bend.

With a partner, discuss what uses these materials might have. Share your ideas with the class.

Did you know?

In the 1930s, brothers Noah and Joseph McVicker invented a putty-like material for cleaning wallpaper. Schoolteachers started using it for craft projects, so the brothers added bright colours and a pleasant scent. This was the beginning of playdough!

Single-use plastics

Plastics are a group of materials that were made just over a century ago from fossil fuels (fuels made from dead plants and animals that take millions of years to form).

Plastics have many benefits for humans. They have improved medicine, made space travel possible, and helped save lives by being used in items like bicycle helmets.

Because most plastics are so easy and cheap to make, many people use them once and throw them away, for example, plastic bags, plastic knives and plastic straws. These are called **single-use plastics**.

Around 40% of all plastic objects made every year are single-use plastics. They are very wasteful and many stay in the environment for hundreds of years.

bags
polystyrene
utensils
cups
straws
balloons
bottles

single-use plastics

Every year, about 8 million tons of plastic waste pollutes the oceans, and harms animals and sea life.

Once in the oceans, the wind, Sun and water break plastics down into **microplastics**. These are small pieces of plastic that are less than 5 mm long.

Because microplastics are so small, they can spread through all the water on the planet.

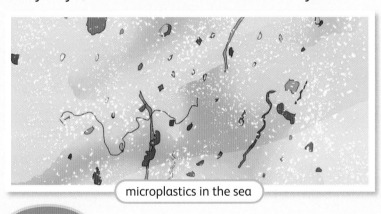

microplastics in the sea

They have been found in every area, including on top of Mount Everest. Recently scientists even found evidence of microplastics in ice cores drilled in the Arctic.

Let's talk

a With a partner, name the different items you use every day that are made from plastic.

b Identify the items on your list that are single-use plastics. Discuss what you think happens to single-use plastics when you have finished with them.

c What can you remember about microplastics and their impact on animals and ecosystems?

d What could you do differently in your daily life to reduce the amount of plastic you use and throw away?

Scientific words
single-use plastics
microplastics

Stop plastic pollution and waste

Think like a scientist!

There are different things we can do to help the environment and stop plastic waste from entering rivers and oceans. We can:

- **Reduce**: Use fewer plastic products. For example, instead of using single-use plastic cups, use ceramic cups that are made from a material that can be used again and again.
- **Reuse**: Use plastic products again or in different ways. For example, turn plastic bottles into sprinklers or animal models.
- **Remove**: Pick up litter.
- **Recycle**: Make plastic waste into something new. For example, make plastic bottles into carpets, paths and benches.

Remember the 4 R's: reduce, reuse, remove and recycle!

1

a Research ideas for reducing, reusing, removing and recycling plastics that could help reduce plastic pollution.

b Create a mind map about stopping plastic pollution, using the ideas that you found. Share your ideas with a partner. Add more ideas to your mind map.

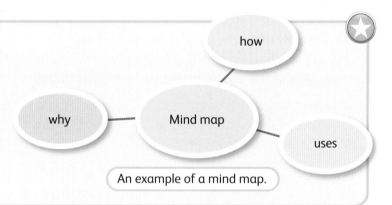

An example of a mind map.

2

a Work in a group. Discuss ideas about organising a campaign to raise awareness about the importance of removing and reducing plastic waste in your school.

b Discuss your ideas with your teacher. Decide on one activity you can try.

c Design a poster to advertise your idea. Try to get your campaign going!

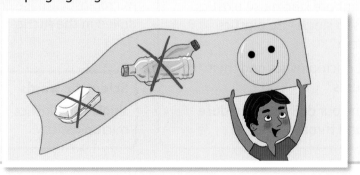

Challenge yourself!

Scientists think that every week we may be swallowing the same amount of plastic that is used to make a credit card. Find out more about microplastics and how they are making their way into our food, drinking water and even the air.

What have you learnt about reversible and irreversible changes?

1

Discuss these questions with a partner. Then share your ideas with the class:

a What do you now know about reversible and irreversible changes that you did not know before?

b What is the most interesting thing you learnt in this unit?

2

Read these descriptions of material changes. Label each change as *reversible* or *irreversible*.

a ice lollies freezing

b concrete setting

c a loaf of bread baking

d chocolate melting

e fruit rotting

f wood burning

3

Draw a mind map, create a cartoon strip or write a story about material changes. Use these words:

reversible   irreversible   physical changes   chemical reactions   melting

condensation   evaporation   freezing   boiling   solution   suspension

What can you do?

You have learnt about reversible and irreversible changes. You can:

✔ say which changes in materials are reversible and which are irreversible.

✔ explain what freezing and melting, and evaporation and condensation are.

✔ link freezing/melting and evaporation/condensation to changes in the state of a substance.

✔ describe and explain the differences between evaporation and boiling.

✔ describe how temperature affects solids dissolving in liquids and relate it to the particle model.

✔ describe the evidence that a chemical reaction has taken place, and state what reactants and products are.

✔ describe how materials chemists use science to help people.

✔ discuss the impact of microplastics and what we can do to stop plastic pollution.

5 Forces

Contact and non-contact forces

What do you remember about about forces?

A force is a push or a pull acting on an object as a result of its interaction with another object. There must be two objects for a force to happen. The two objects can either be in contact (**contact forces**), or at a distance from each other (**non-contact forces**).

Scientific words

contact forces

non-contact forces   motion

Let's talk

a With a partner, discuss the names of all the different forces you can remember.

b Now group all the forces you have named into contact and non-contact forces.

c What can you remember about force arrows?

1

The overall forces acting on an object will affect its **motion** (the way it moves).

Look at the diagrams below.

Predict what you think will happen to the motion of the object in each diagram. Explain why you think so for each one.

A

B
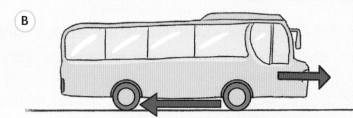

C

D
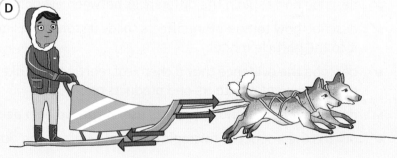

Force diagram arrows

Think like a scientist!

Force diagrams show the forces that act on an object. Force diagrams use arrows to represent forces. The force arrows must meet these four rules:

- They must show the direction in which the force is acting. This is shown by the direction the arrow is pointing.

- They must show how strong each force is. The longer the arrow, the stronger the force.

- They must show where the force is acting from. The base of the arrow (the flat end), shows us where the force is acting from.

- They must be labelled with the name of the force, including the two objects that make up this force.

Look at this example:

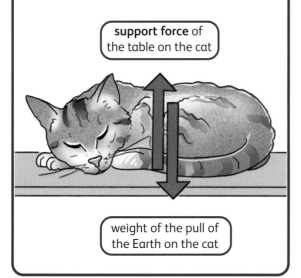

support force of the table on the cat

weight of the pull of the Earth on the cat

Scientific words

force diagram
support force

Let's talk

a Discuss with a partner how the example in the *Think like a scientist!* box meets the four rules for force arrows.

b What do the force arrows show about the cat's motion?

1

Some Class 6 learners produced the diagrams below to show forces acting in different situations. Improve the diagrams so that they meet the four rules for force arrows.

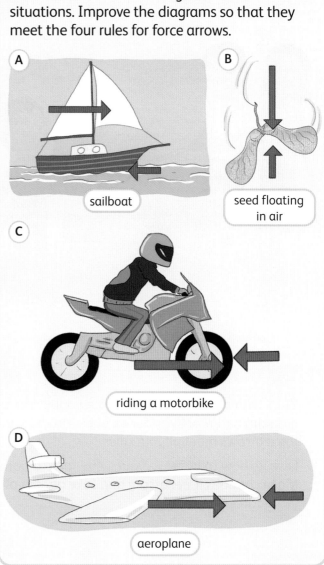

A

sailboat

B

seed floating in air

C

riding a motorbike

D

aeroplane

Modelling forces

1

You will need...
- card to cut up
- scissors

a Work in a group. For each picture, choose one person to act as the object.

b Draw, cut out and label arrows. Place them in the correct positions on the pictures to show how the forces are acting on the object.

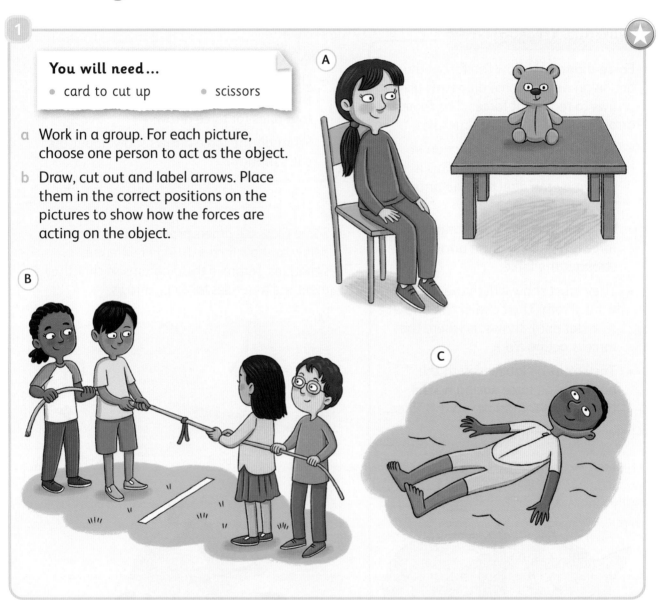

A

B

C

2

a Work with another group. Explain your living force diagrams from Activity 1. Ask questions to see if your force arrows meet all four rules.

b Discuss how to improve your living force diagram. Make changes if necessary and take photos of your living force diagrams.

3

Use the photographs of your final living force diagrams, produce an instructional leaflet or video to help Class 6 learners develop their use of force arrows.

Motion and balanced forces

Think like a scientist!

Forces can affect motion. Motion is the way an object moves. If forces are balanced, it means that the forces are the same strength but are acting in opposite directions.

Balanced forces do NOT change the motion of the object. This does not mean that if forces are balanced the object cannot be moving.

If the object is moving, and there are balanced forces acting on it, the object will continue to move at the same rate and in the same direction.

If the object is still, it will stay still.

1

The only force acting on the panda is the downward pull of weight from the Earth.

The panda is not moving. There are no forces acting on it.

There are two forces acting on the panda: the weight from the pull of the Earth and the support force of the table.

A table cannot push. It is just in the way of the panda.

a The panda is an example of balanced forces. How do you know?

b Who do you think has the correct scientific answer?

c Why do you think the other answers are incorrect?

d What would you do to convince others about the correct answer?

Scientific word
balanced

2

You will need...
- track (two rulers taped to a board with a gap between them for the ball to run down)
- small ball or marble
- measuring tape
- timer or stopwatch

a Work in a group. Discuss how you can make the ball move down the track so that its motion stays the same throughout?

b How will you know if the ball's motion is the same throughout? What measurements will you take to prove this?

c Produce a results table and record your measurements. If possible, take a video of the experiment to show when the ball is moving with the same motion throughout.

d Draw a force diagram of the ball when it is moving with balanced forces. Describe this motion.

Challenge yourself!

If a rocket is moving in space, will you need to use the boosters to keep it moving? Explain your thinking. Do research to find out when and why astronauts use the rocket's boosters.

Motion and unbalanced forces

Think like a scientist!

If the forces acting on an object are balanced, it will either stay still or it will keep moving at the same rate. If it is moving and the motion is not changing, we call this **steady speed**.

If forces are **unbalanced**, it means that the force acting in one direction is greater than the force acting in the opposite direction. Unbalanced forces DO change the motion of the object.

If the object is moving, unbalanced forces will make it change speed or direction.

Let's talk

Discuss with a partner what will happen if one learner pulls another learner sitting on a trolley with the same force all the time. Explain your reasons.

Work safely!

Do not stand up on the trolley.

Think like a scientist!

When the forces acting on an object are unbalanced, the motion of the object will always change. The motion of the object will either continually increase (speed up), or it will continually decrease (slow down).

Scientific word

steady speed   unbalanced

1 **You will need...**
- trolley
- rope

a   You will investigate how unbalanced forces change the motion of an object.

b   What happens to the learners pulling the trolley? (**Hint**: They must change the way they move!).

c   What measurements can you take to prove who has the strongest pulling force? Take and record the measurements to see.

d   How would you describe the motion of the learners pulling the trolley?

2

a   Look at the force diagrams on page 68. Which show balanced forces? Which show unbalanced forces? How do you know?

b   How would you describe the motion at the moment in time shown in each example on page 68?

c   Draw a force diagram where forces are unbalanced, and an object is slowing down.

An interview with a toyologist

What do you do?

I use my scientific understanding of how forces make things move and their mechanical make-up to develop and make toys more lifelike.

How did you become a toyologist?

I studied mechanical engineering and enjoyed the different sides of design. So, I decided that I wanted to be a toy designer.

What do you create?

I am working on creating human-like facial features in dolls. I am creating doll's eyes that work like human eyes so that they look real.

How does this work?

I use a technology called nanomuscles. These are very thin threads that conduct electrical currents. When the currents run through the threads, they make them stretch or contract. I can then use this change in motion to move the eyes of the doll.

What helped you make the eyes so lifelike?

I spent a lot of time observing how human eyes work, especially the eyes of babies. I used this to make doll's eyes that work in the same way. I tested the dolls by asking young children to play with them and telling me what they thought.

What are you excited about next?

I think by combining science and creativity, we will come up with toys that we haven't even imagined. I just can't wait.

Toyologists must think about the different materials they use and how smoothly the eyes move. If there is too much friction, the eyes will not appear to move like those of a human.

1

You will now be a toyologist.

a Design an action toy that is made for a specific sporting activity.

b You will need to draw a design for your action toy, including the accessories it needs to use.

c You need to explain how you have used ideas related to forces in the design of your toy.

Mass and weight

Think like a scientist!

People often confuse **mass** and **weight**. The mass of an object is a measure of the amount of matter it contains. The more matter an object contains, the greater its mass. Mass is measured in kilograms (kg) or grams (g).

Weight is a downward force. Weight happens when two masses interact and attract each other. We call the attraction between the two masses **gravity**. Weight is the force experienced by the mass that is being pulled down by gravity.

Weight is an example of a non-contact force because the objects can be at a distance from each other. As weight is a force, it is measured in newtons (N).

Let's talk

Four children are standing at different points on the Earth. They are all about to drop an apple. Discuss which way you think each apple will fall. Explain your thinking. Draw a picture if it helps you.

1

You will need…

- scale that measures mass in kilograms
- force meters that measure weight in newtons
- objects to measure

Work in a group to investigate the relationship between mass and weight.

a Collect ten different objects from the classroom to measure.

b Copy this table. Write the names of the objects you collected in the first column.

Object	Mass in kg	Weight in N	Ratio of weight to mass

c Use the scale to measure the mass of each object in kilograms. Record the mass.

d Use a force meter to measure the weight of each object in newtons. Record the weight.

e Compare your results with those of another group. What similarities are there?

f Is there a pattern linking the mass of an object to its weight? If there is, describe it.

g Find out the ratio of the weight divided by the mass for each object. Are the results what you expected? Can you see a pattern? Explain the pattern.

Scientific words

mass   weight   gravity

Let's talk

a What type of scientific enquiry is Activity 1?

b What features of the enquiry make you think this?

Walking on the Moon

Think like a scientist!

Gravitational attraction is an interaction that happens between masses throughout the universe. The gravitational attraction between the masses pulls on an object and creates its weight. The Earth is a very large mass and it pulls down on every object on the planet.

The last column of the table that you completed in Activity 1 on page 74 tells you the gravitational attraction the Earth's mass has on every kilogram on our planet. Gravitational attraction is measured in newtons per kilogram (N/kg).

This means, you now know that every 1 kilogram of mass on Earth weighs about 10 **newtons**. But this is not true everywhere in the universe. It is only true here on Earth!

For example, an astronaut has a mass of 60 kg. On Earth, her weight is about 600 N.

The Moon's gravitational attraction is about one-sixth of the Earth's gravitational attraction. Therefore, on the Moon, the astronaut weighs about one-sixth of what she weighs on Earth. This means that she weighs about 100 N. Her mass is still 60 kg.

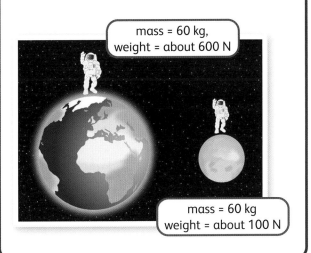

mass = 60 kg,
weight = about 600 N

mass = 60 kg
weight = about 100 N

1

Calculate the weight of each of the following objects on the Earth and on the Moon.

The first one has been done for you.

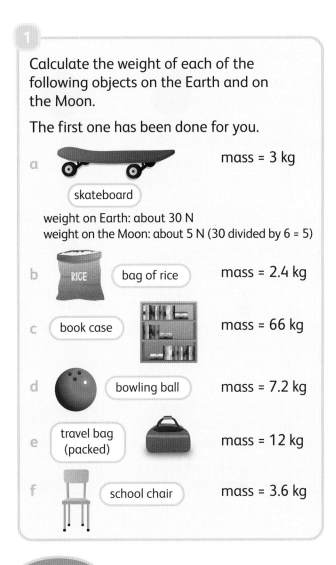

a skateboard mass = 3 kg

weight on Earth: about 30 N
weight on the Moon: about 5 N (30 divided by 6 = 5)

b bag of rice mass = 2.4 kg

c book case mass = 66 kg

d bowling ball mass = 7.2 kg

e travel bag (packed) mass = 12 kg

f school chair mass = 3.6 kg

Let's talk

Discuss these questions with a partner:

a The astronaut's weight on the Moon is less than her weight on Earth. Why?

b The astronaut's mass on the Moon is the same as her mass on Earth. Why?

c Share your ideas with the class.

Scientific words

gravitational attraction   newtons

Do gases have mass?

Let's talk

Discuss these questions with a partner:

a Do gases have mass?

b What do you think will happen if one of the balloons in the picture is popped? Explain your reasons.

1

You will need...

- long stick or metre ruler
- three pieces of string about 30 cm long
- two balloons of equal size
- balloon pump (you can share this with other groups)

Work in a group.

a Blow up the two balloons until they are equal in size and tie them off. If this is difficult, ask the teacher for help.

b Attach a piece of string to each balloon. Tie each balloon to opposite ends of the ruler. The balloons should now dangle below the ruler, just like in the diagram above.

c Tie the third string to the middle of the ruler and hang it from the edge of a table.

d Adjust the middle string until everything balances and the ruler is horizontal.

e What is inside each balloon? How do you know that the forces are balanced? Explain using ideas you have learnt about forces.

Work safely!

Be careful when popping the balloons. The parts of a burst balloon can fly off very quickly, so stand a safe distance away.

Think like a scientist!

On the previous page, you learnt that all objects have mass, even those that are too small for us to see. This includes gases and the air around us.

2

a Now you have set up the equipment and have decided what you think will happen, pop one of the balloons.

b What happens? Was your prediction correct?

c Can you explain what happened using ideas linked to forces?

3

a What do you think would happen if you popped the other balloon? What are your reasons?

b Pop the second balloon. What happens? Was your prediction correct?

c Can you explain what happened using ideas linked to forces?

d Produce a cartoon strip to explain to others what you did, and what you have learnt about gases and mass.

Did you know?

The total mass of Earth's atmosphere is about 5.7 quadrillion tons. That's about the same as 570,000,000,000,000 adult Indian elephants!

Science in context

Under-inflated tyres

We use air to **inflate** many different things including tyres on vehicles.

If a tyre is not inflated properly, there can be problems. Over-inflated tyres can result in tyres wearing out too soon, losing grip on the road, or creating a bumpier ride. Under-inflated tyres can also cause problems.

Under-inflated tyres have been classed by some as a global issue. This is because tyres that are under-inflated increase:

- the money people spend on fuel
- the amount of carbon dioxide produced by vehicles
- the distance required to stop
- difficulties in steering a vehicle precisely
- the possibility of a vehicle **aquaplaning**
- tread wear, making a tyre more likely to get damaged and deflate rapidly.

Scientific words

inflate
aquaplaning

under-inflated | properly inflated | overly-inflated

1

Work in a group.

a Use your research skills to find out about the different issues related to under-inflated tyres. How do under-inflated tyres contribute to these issues? How are the issues reduced if people use properly inflated tyres?

b Present your research in an interesting way to the rest of your group.

c Take turns to sit in the 'hot seat' and answer questions from the rest of your class about the issues you have investigated.

Did you know?

You should inflate your bicycle tyres at least once or twice a week, or before every ride if you don't go out that often. Road bike tyres have been known to lose pressure after 4–5 days of standing still.

When did you last inflate your bicycle tyres?

Weight in water

Think like a scientist!

You have learnt that all objects have a mass, and that all masses can apply a force on other objects. This means that solids, liquids and gases can apply a support force on other objects they interact with.

You have also learnt that a liquid applies a force on an object when the object interacts with the liquid.

If a liquid can push back enough on an object, it will float. If a liquid cannot push back enough on the object, it will sink.

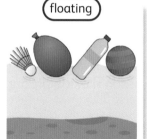
floating

sinking

Let's talk ⭐

This brick feels heavy.

This brick feels lighter now!

a Do objects weigh less in water than in air? Explain your thinking to a partner.

b Draw a force diagram to explain your ideas.

1 ⭐

You will need...

- four film canisters, or similar plastic containers with lids
- sand
- plastic tank
- water
- pen to label canisters

You are going to investigate how the mass of an object affects how well it floats or sinks. Work in a group.

a Prepare and label four canisters:

- full of sand
- two-thirds full of sand
- one-third full of sand
- no sand.

b Predict what you think will happen when you place all four canisters into a plastic tank full of water. Explain your reasons.

c Place all four canisters into the plastic tank. What do you see? Make a note of your observations.

d Explain, using ideas about forces in liquids, what you observed. Include a force diagram for each canister in your explanation.

Investigating floating and sinking

You will need...
- aluminium foil
- scale that measures mass in grams
- a number of small 10 g masses (10 g is 0.01 kg)
- bucket
- water

You are now going to investigate how the shape of an object can affect how well it floats or sinks. Work in a group.

a Make four boats using the aluminium foil. Your boats must use the same amount of aluminium foil, but each must be a different shape.

b Predict which boat you think will hold the most mass.

c Copy this table.

Boat shape	Mass (kg)
1	

d Place one boat at a time on top of the water in the bucket. Carefully add one mass at a time, counting them until the boat sinks. Make a note of the amount of mass the boat held before sinking. Complete your table.

e Do the same for each of the different shaped boats.

f Explain your results using ideas about forces in liquids. Include force diagrams for each of the different boats in your explanation.

Scientific word
upthrust

Think like a scientist!

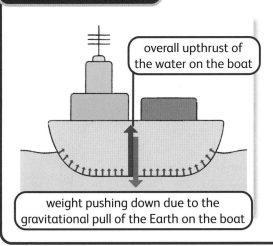

overall upthrust of the water on the boat

weight pushing down due to the gravitational pull of the Earth on the boat

The name of the force that pushes up on objects in water is **upthrust**.

As the mass of an object increases, the weight of the object increases. If the weight of the object is greater than the force of upthrust, the object will sink.

The shape of an object can change the size of the upthrust force. If the surface area of the mass placed in the water is increased, the size of the upthrust force can be increased.

This is how metal boats containing lots of cargo are able to float on water.

What have you learnt about forces and motion?

Let's talk

Explain each word/phrase to a partner:

a contact force   b non-contact force   c mass

d weight   e friction   f upthrust

2

Work with a partner to create a mind map showing what you know about forces and motion. Try and use all of these words.

> contact forces   upthrust
>
> non-contact forces
>
> mass   force diagram
>
> balanced forces   weight
>
> unbalanced forces
>
> floating   motion
>
> support force   friction
>
> air resistance   sinking
>
> gravitational attraction

1

Draw a force diagram for each of the following pictures. For each diagram, make sure:

- your arrows meet the four standards for a force arrow
- you say whether the forces are balanced or unbalanced
- you describe the motion of the object.

a child pulling a sledge with a constant force

a child sitting still on a swing

a raindrop falling

What can you do?

You have learnt about forces and motion. You can:

✔ use force diagrams to show the name, size and direction of forces acting on objects.

✔ describe the effect of forces on an object at rest.

✔ describe the effect of forces on an object that is moving.

✔ describe the difference between mass, measured in kilograms (kg), and weight, measured in newtons (N).

✔ explain how the weight of an object changes but the mass does not, and link this idea to gravitational attraction between masses.

✔ explain how the mass and shape of an object can affect if it floats or sinks.

Mains electricity and cells

What do you remember about using electrical power safely?

Electrical power comes from these main sources: **mains electric sockets**, **cells** and batteries. Remember: a cell is a single unit; a battery usually contains two or more cells.

These sources of electrical power can be dangerous, so you need to know how to keep yourself and others safe. What can you remember about using electrical power safely?

cells and batteries

electric sockets connect to the mains

1

For each picture, write a rule for using electrical appliances and mains sockets safely. The first one has been done for you.

A

Do not put drinks or liquids near electrical appliances.

B

C

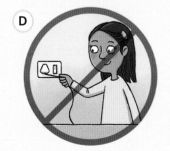

D

E

2

a Which pictures show a cell being used safely?

b Explain why each of the other ways of using a cell is unsafe.

A   B

C   D

Let's talk

Which source of electrical power is more dangerous – mains electric sockets, cells or batteries? Why? Discuss your ideas with a partner.

Scientific words
mains electric sockets   cells

Electrical circuits

Think like a scientist!

Electrical conductivity is how well a material conducts electricity. A complete electrical loop (with no gaps) is needed for any electrical **component** to work. We call this complete electrical loop a **circuit**.

Electrical circuits will not work if there are breaks anywhere in the loop. This means that for circuits to work:

- there needs to be a source of electrical power (cell or mains)
- all the components must be connected correctly
- there must be no gaps in the circuit.

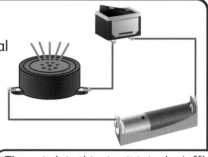

The switch in this circuit is in the 'off' position. Once switched on, this is now a complete electrical circuit. You can put your finger anywhere on the circuit and trace all the way around the loop.

Scientific words

electrical conductivity   component   circuit

1 You will need...

- circuit components

Look at these circuit pictures. Which circuit is the odd one out? Explain why you think so.

a   Predict in which circuit the buzzer will sound.

b   Explain why the buzzer will not sound in each of the other circuits.

c   Test your predictions by tracing your finger around the electrical loop, and then making the circuits. Were your predictions correct?

d   Write instructions for someone else to follow, explaining how to make a circuit that will make a buzzer sound.

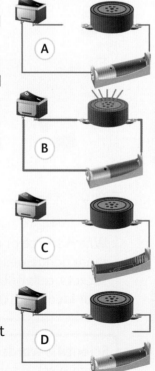

Let's talk

Some Class 6 learners were discussing complete circuits and electrical devices that plug into the mains.

Arnita: I think there is only one wire, so it cannot be a complete circuit and the table lamp will not light.

Bipasha: I think you only need one wire from the power source to the component, so the table lamp will light.

Eva: I think that the wire that goes from the mains to the table lamp has more than one wire inside it to make a complete circuit, so it will light.

What do you think? Explain to a partner who you agree with and why.

Circuit symbols

Think like a scientist!

Each component of a circuit can be drawn as a circuit symbol. Here are some of the standard **circuit symbols** used throughout the world:

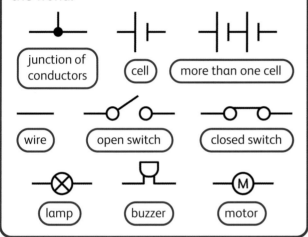

junction of conductors   cell   more than one cell

wire   open switch   closed switch

lamp   buzzer   motor

Let's talk

Discuss these questions with a partner:

a Why are symbols used to represent the components in a circuit?

b Why are the symbols standard throughout the world?

c Share your ideas with the class.

2

a Make nine blank cards.

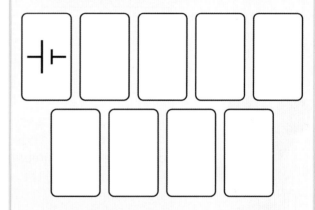

b Draw a different circuit symbol on the front of each card.

c Write the name of the circuit symbol on the back of each card.

d Show the circuit symbols to your partner one at a time. Time how long it takes them to name all the symbols correctly.

e Show the names of circuit symbols to your partner one at a time. Time how long it takes them to draw all the symbols correctly.

f Over time, repeat the activity and try and beat your personal best times.

1

Draw the correct symbol for each component.

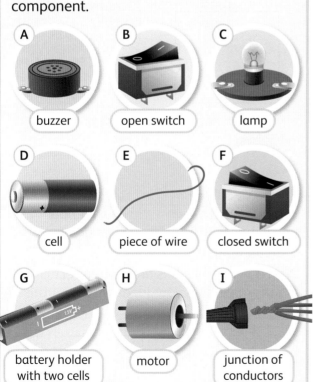

A buzzer
B open switch
C lamp

D cell
E piece of wire
F closed switch

G battery holder with two cells
H motor
I junction of conductors

Scientific word
circuit symbols

Circuit diagrams

Think like a scientist!

A **circuit diagram** shows the components in a circuit and how they are connected to one another. Circuit diagrams use the standard circuit symbols. Here is a circuit:

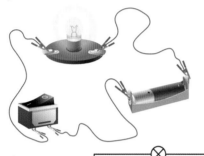

Here is the same circuit drawn as a circuit diagram.

Is the circuit a complete circuit? Place your finger anywhere on the circuit. Can you trace it all the way around?

Let's talk

Look at the pictures of the real circuit and the circuit diagram in the *Think like a scientist!* box. Discuss these questions with a partner:

a What similarities are there between the real circuit and the circuit diagram?

b What differences are there?

c Which of the two pictures (real circuit or a circuit diagram) would learners in another country be able to copy more easily? Why?

Scientific word
circuit diagram

1

Match each circuit with the correct circuit diagram.

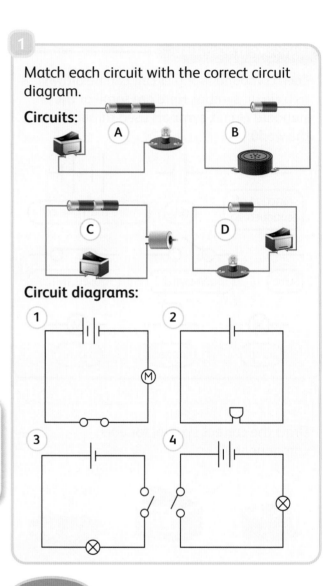

Let's talk

Three Class 6 learners all used the same components to build a circuit that had two lamps in it. Then they each drew a circuit diagram.

Who do you think has drawn a correct circuit diagram? Discuss your ideas with a partner.

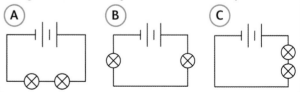

Creating circuits and using circuit diagrams

Think like a scientist!

Sometimes you need to be able to control the components in an electrical circuit. For example, in a fridge you want the lamp to come on when you open the door to see inside and you want the lamp to go off when you shut the door.

To do this, we use switches that help us make a gap and break the electrical circuit.

You will need...

- circuit components
- aluminium foil
- wire
- scissors
- stiff card
- clear tape
- paper clips

a Design and make a circuit. It can be either:

- a burglar alarm for the classroom door, so a bell rings if someone opens the door

- a pressure pad alarm, so a bell rings if someone sits on the teacher's chair.

b Draw and label a picture to explain how your circuit works. Make sure you show how your design uses a switch to open or close a gap in the electrical loop to make a complete circuit.

c Draw a circuit diagram to explain how your circuit works. Give it to someone else to evaluate your design.

d Create a flip book, using pictures and key words, to show others how to build your alarm system.

Let's talk

a Explain to a partner how your device works.

b Discuss whether it matters where you put the switch in the circuit.

Electric vehicles and global warming

One of the biggest challenges facing the human population, and the planet, is **global warming**.

Global warming happens when greenhouse gases build up in the atmosphere. The **greenhouse gases** stop energy from escaping into space and make the planet warmer. This is called the **greenhouse effect**.

Human activities are the main causes of global warming. One of the biggest ways we are increasing greenhouse gases is through transportation. Greenhouse gas emissions from transportation come from burning fossil fuels (including petrol and diesel) for our cars, trucks, ships, trains, and planes.

Scientists and engineers around the world are looking to develop electric vehicles to help lower the amounts of greenhouse gases made. These electric vehicles can be charged using **renewable** electricity. This is electricity made by using solar, water or wind power, rather than by burning **fossil fuels** (including coal and gas). Electric vehicles could also improve public health. They cause less air pollution, which could lead to a decrease in diseases like asthma and lung cancer.

Let's talk

Discuss these questions with your partner:

a Do you know anyone who uses an electric car?

b Do you think you will use an electric car when you grow up? Explain your reasons.

1

Do research to find out:

a the names of the gases that are classed as greenhouse gases

b which human activities produce greenhouse gases

c different ways that we can help reduce the amount of greenhouse gases made.

Scientific words

global warming   greenhouse gases
greenhouse effect   renewable
fossil fuels

2

You are going to do an investigation to decide if carbon dioxide is a greenhouse gas that is contributing to global warming.

You will need...

- two jars with lids
- two thermometers
- jug of water
- some vinegar
- bicarbonate of soda
- teaspoon
- pen to label jars
- timer or stopwatch

a Pour equal amounts of water and vinegar into both jars.

b Place a thermometer in each jar. If the thermometer is too long for the jar, you can make a hole in the metal lid so it fits (you might need your teacher to help you with this). Or you can make a lid out of card and make a hole in it for the thermometer to stick through.

c In one jar only, add a teaspoon of bicarbonate of soda to produce carbon dioxide. Label this jar *carbon dioxide*. Label the other jar *no carbon dioxide*.

d Place both jars in a sunny spot. Measure and record the readings on the thermometer every 5 minutes for an hour.

e Place both jars in a shady spot. Measure and record the readings on the thermometer every 5 minutes for an hour.

f Draw a graph of your readings. What does the graph show you?

g What conclusions can you make about carbon dioxide as a greenhouse gas? Does it add to problems related to global warming or not? If it does, how does it? What is your evidence to support your view?

3

From the research you carried out in Activity 1 and the investigation you did in Activity 2, produce a leaflet that advertises problems with greenhouse gases and the different ways that we can work to reduce global warming.

Work safely!

Be careful when using scissors to make a hole in the lid.

Did you know?

Nine out of ten road deaths are caused by human error; that is more than 1 million people killed each year. To make roads safer, scientists and engineers are now developing electric cars that will be self-driven (have no human driver).

Let's talk

Discuss these questions with your partner:

a What Positive, Minus and Interesting (PMI) reasons can you think of why we should develop driverless cars?

b How many PMI reasons can you come up with?

How do we know about electrical circuits?

We know about electrical circuits because of the work of many scientists over hundreds of years.

These scientists asked questions about things they observed and the world around them. To find out the answers to their questions, they carried out investigations. In these investigations, they collected evidence by making observations and taking measurements. They used creative thinking to come up with new ideas that would explain these observations and measurements.

1

Match each scientist with the correct discovery or invention. Do some research to find out the answers.

Luigi Galvani

Alessandro Volta

A Carried out a dangerous experiment that proved that lightning and tiny electric sparks were the same thing, by flying a kite in a thunderstorm.

B Discovered that when he touched a dead frog's leg with a knife, it twitched violently.

Benjamin Franklin

C Invented an early type of electric cell called a voltaic pile.

2

a In pairs, choose another scientist linked to electricity, one that isn't mentioned in Activity 1.

b Use your research skills to find out more about your chosen scientist. Find out:
- the dates of their birth and death
- the country they were from
- how they contributed to scientific knowledge about electrical circuits
- any other inventions or discoveries they made
- anything else about the scientist that you find interesting.

c Present what you have found out in an interesting way.

You could:
- role-play the scientist and tell 'your' life story
- present a PowerPoint slideshow
- give a short talk, followed by a quiz.

Series circuits

Think like a scientist!

If there is only one electrical loop that can be traced with your finger, this is called a **series circuit**. In a series circuit, all the components are connected in the same loop.

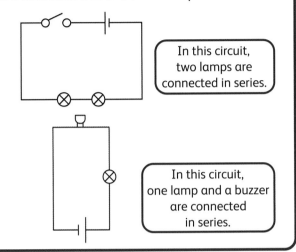

In this circuit, two lamps are connected in series.

In this circuit, one lamp and a buzzer are connected in series.

Scientific word
series circuit

2

You will need...
- circuit components

a Make the circuits in Activity 1 and record your observations.

b Are the predictions you made supported by the evidence you have collected?

c What conclusions can you draw from the evidence you have collected?

3

Describe three ways to change this circuit to make the motor:

a spin more quickly

b spin more slowly.

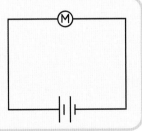

1

Look at the circuit diagrams below. Each diagram shows a change being made in a series circuit:

- The diagram to the left of the arrow shows the circuit before the change.
- The diagram to the right of the arrow shows the circuit after the change.

Predict what will happen when each circuit is changed.

a Predict what will happen to the brightness of the lamps.

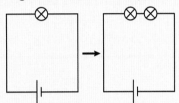

b Predict what will happen to the brightness of the lamps.

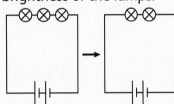

c Predict what will happen to the brightness of the lamp and the speed of the motor.

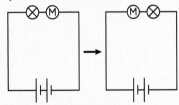

d Predict what will happen to the volume of the buzzer.

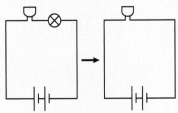

Parallel circuits

Think like a scientist!

If components are connected in more than one loop, this is called a **parallel circuit**.

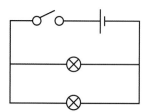 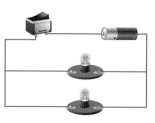 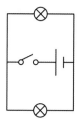 

In this circuit, two lamps are connected in parallel.

This is what the parallel circuit in the diagram looks like in real life.

Two lamps are also connected in parallel in this circuit.

This is what the parallel circuit in the diagram looks like in real life.

Even though they look different, both of these electrical circuits have two lamps in parallel. This is because a complete electrical circuit has to include the cell and the lamp. If you trace the loops with your finger, you can see that each lamp is in a separate loop.

1

You will need...

- circuit components

Some learners have found that when they put two or more components in a circuit, it does not always work. Here is a different way to make your circuit.

It is called a parallel circuit.

Make the circuit in the circuit diagrams in the *Think like a scientist!* box above.

Scientific word

parallel circuit

2

You will need...

- circuit components

Each circuit diagram below shows a change being made in an electrical circuit:

- The diagram to the left of the arrow shows the components connected in series.
- The diagram to the right of the arrow shows the components connected in parallel.

a Make the circuits. Observe what happens when you change how the components are connected.

b Describe what happens to the brightness of the lamps.

c Describe what happens to the speed of the motors.

d Describe what happens to the volume of the buzzers.

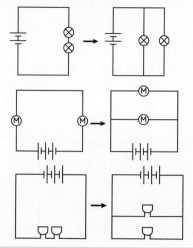

Building devices using series and parallel circuits

Let's talk

Emily has made a set of model traffic lights. She used a parallel circuit in her model. Below is a circuit diagram of the circuit she used.

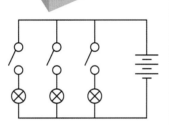

a Match the components in the circuit diagram with the components in the picture of her model.

b How many electrical loops are there in this circuit?

c Explain why Emily had to use a parallel circuit for her model traffic lights, rather than a series circuit.

Let's talk

Discuss these questions with a partner:

a Which electrical device will you make?

b What components will you include in the circuit? Why?

c What sort of circuit will you need to build – a series circuit or a parallel circuit?

1 **You will need...**
- equipment for making your choice of electrical device

Design one of these electrical devices:
- a lighthouse, with two lamps, so it can be seen a long way out at sea
- a set of model traffic lights, controlled by switches
- a model house with lights, with separate switches in each room
- a model vehicle, driven by an electric motor
- a child's bedside lamp that includes a dimmer switch
- a fish tank pump and light system. The motor for the pump must always be on, with a switch used to turn the lamp on at night.

2

a Draw and label a diagram to show what your electrical device will look like.

b Draw a circuit diagram to show the electrical circuit you will use in your device.

c Write a list of the equipment you will need.

d Make and test the device you have designed.

3

When you have made and tested your device, answer these questions:

a Did you have any problems making your device? If so, how did you solve them?

b Did you make any changes to your design as you went along? If you did, what were the changes? Why did you make them?

c Is your finished device different from your drawing? If it is, explain how it is different. Why did you have to make changes?

What have you learnt about electrical circuits?

1

Look at these circuit diagrams. Each diagram shows a change being made in an electrical circuit.

The diagram to the left of the arrow shows what the circuit is like before the change.

The diagram to the right of the arrow shows the circuit after the change.

Predict what will happen when each circuit is changed.

a Predict what will happen to the speed of the motor.

b Predict what will happen to the volume of the buzzer.

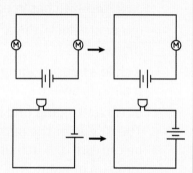

2

Draw a circuit diagram to represent each of these circuits.

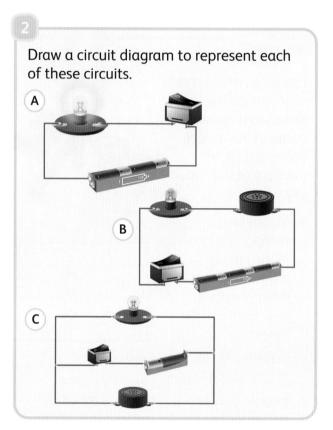

A

B

C

> **Let's talk**

a Explain the predictions you made in Activity 1 to a partner.

b Discuss which parts of this unit you both found the most difficult to understand.

c How can you help each other get better at the parts you identified?

What can you do?

You have learnt about electrical circuits. You can:

✔ describe what is needed for an electrical loop to work.

✔ explain how switches work.

✔ use diagrams, and conventional symbols, to draw series and parallel circuit diagrams.

✔ predict and test the effects of making changes to circuits, by adding or taking away components or electrical loops.

✔ use your ideas about circuits to design devices.

How we see things

We see when **light** enters our eyes, the **light detectors** in our bodies. Even though light is all around us, we do not really see the **light rays**. We really see only two types of objects:

- objects that are **light sources** giving off their own light
- objects that **reflect** light, which is then detected by our eyes.

Light can travel through **transparent** materials, and cannot travel through **opaque** materials. Some **translucent** materials let some light through, but we cannot see objects clearly through them. Most objects we see are opaque.

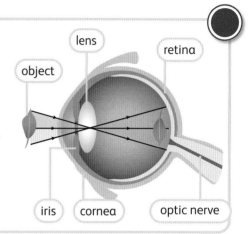

lens · retina · object · iris · cornea · optic nerve

Let's talk

Discuss these questions with a partner:

a How many different light sources can you remember?

b How many different light detectors can you name?

c Some Class 6 learners named 5 transparent, 4 translucent and 7 opaque materials. Try to beat their record.

Scientific words

light	light detectors
light rays	light sources
reflect	transparant
opaque	translucent

1

Some Class 6 learners are discussing how they see a flower in a vase on the table and have produced light ray diagrams to explain their ideas.

Which of the following diagrams do you think best shows what is happening? Explain your reasoning.

What do you think is wrong with each of the other diagrams?

Amir

Kim

Lucy

Harish

When light meets opaque objects: reflection

Think like a scientist!

All opaque objects reflect light, even black ones. However, some opaque objects reflect light in a special way. They allow you to see the image of the object in them. These clear images that can be seen are called **reflections**.

In the picture on the right, you can see mountains reflected in the lake. The surface of the lake is reflecting light in a special way to make the reflections. The mountains are the objects, and the reflection is the image of the mountain.

When rays of light hit (shine on) most opaque surfaces, the rays are reflected at lots of different **angles** because the surfaces are uneven and rough. Rough surfaces do not usually make clear images. This is called **diffuse** reflection.

When rays of light hit a very smooth opaque surface, such as a calm lake, all the rays reflect at the same angle. This produces (forms) reflections and is called **specular** reflection.

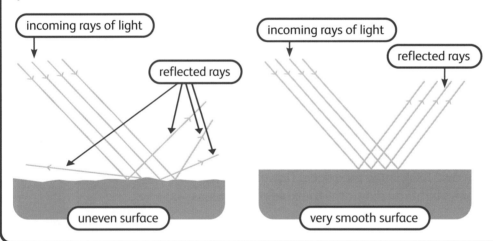

incoming rays of light

reflected rays

uneven surface

incoming rays of light

reflected rays

very smooth surface

Where have you seen reflections?

1

A **mirror** produces reflections because all the light that shines on it bounces back off in the same direction. Mirrors are used in many places.

Discuss this question with a partner: What job does each of these mirrors do?

- Rear-view mirror in a car
- Dental mirror
- Baby toy with a small mirror
- Security mirror (such as in a shop)
- Mirror wall in a dance studio

Scientific words
reflections
angles
diffuse
specular
mirror

Investigating reflections

Think like a scientist!

The reflection of an object is its **mirror image**. If you compare an object to its mirror image, you will notice that the mirror image is in reverse (back to front). For example, if you look at the letter *r* in a mirror, you will see it in reverse.

Mirror writing is writing that uses mirror images of letters.

1

Aanya wrote a secret message to her friend using mirror writing. She held a mirror next to a sheet of paper. Then she wrote her message on the paper. She formed the letters so that they were the right way around in the mirror.

a When Aanya had finished writing her secret message it looked like this. What does the message say?

2 **You will need…**
* a mirror

a Write your own secret message using mirror writing.

b Swap your message with someone else and try and read their message using your mirror.

Did you know?

There is a mirror on the Moon. Astronauts left it there in the 1970s. Scientists aim laser beams at the mirror from telescopes on Earth. They measure the time it takes for the laser light to travel to the Moon, hit the mirror and reflect back to the Earth. They can then calculate the exact distance to the Moon.

Scientific word
mirror image

b Here is the reply. What does it say?

I like orange juice the best.

Reflecting light rays

1

The diagram shows a light ray reflecting from a mirror.

A 'normal' line is drawn from where the light hits the mirror at 90° to the surface. (A 'normal' line is drawn at right angles to the surface.)

Angle *a* is the angle between the normal and the incoming light ray. This is called the **angle of incidence**.

Angle *b* is the angle between the normal and the reflected light ray. This is called the **angle of reflection**.

If angle *a* gets bigger, what happens to angle *b*?

If angle *a* gets smaller, what happens to angle *b*?

Discuss your ideas with a partner.

Scientific words

angle of incidence
angle of reflection

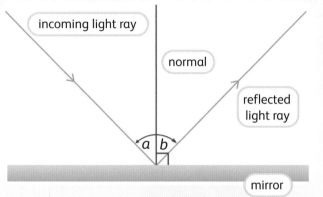

incoming light ray

normal

reflected light ray

a *b*

mirror

2

You will need...

- flashlight
- card
- scissors
- sticky tape
- table
- flat mirror
- protractor

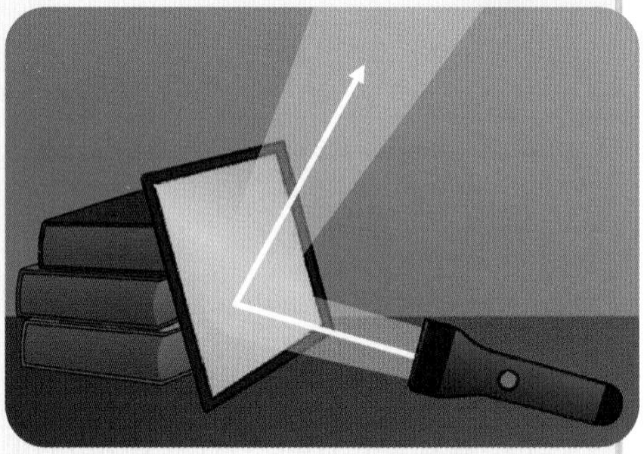

You are going to investigate how the angle at which a light ray is reflected from a mirror is related to the angle at which it hits the mirror.

a   Cut out a circle of card to cover the front of the flashlight.

b   Cut a narrow slit in the circle of card.

c   Tape the circle of card with the slit in it over the front of the flashlight.

d   Place a mirror on its edge on a table. Prop it up against something.

e   Darken the room as much as you can.

f   Lay the flashlight on the table and switch it on.

g   Shine the flashlight onto the mirror at an angle.

h   Use the protractor to measure and record the angle of the incoming light ray and the angle of the reflected light ray.

i   Shine the flashlight onto the mirror at a different angle. Measure the angles again.

What did you find out from doing the investigation in Activity 2? Discuss your ideas in a group.

Periscopes

Think like a scientist!

You have seen that light always reflects from a mirror at the same angle that it hits the mirror.

A **periscope** lets you see places you might not otherwise be able to see. Examples are over the top of walls or fences, or around corners. Periscopes were first used in submarines so that sailors could see above the water.

A simple periscope is a long tube with a flat mirror at each end. The mirrors are at an angle of exactly 45°. Light rays hit the top mirror at 45° and reflect off it at the same angle. The light then travels down to the bottom mirror. When the light hits the bottom mirror, it is again reflected at 45°, and into the eyepiece.

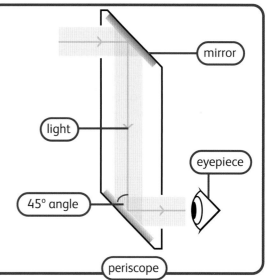

Let's talk

Look at this homemade periscope. Discuss these questions with a partner:

a How do you think the periscope was constructed (made)?

b Why do you think the people are using periscopes?

Scientific word

periscope

1

Use angled mirrors to send a beam of light from a flashlight:

• around a corner

• over the top of an obstacle.

2

a Design and make your own periscope.

b Use your periscope to look around your classroom.

c Use the *Scientific words* you have learnt, to help you do the following:

• Describe what you did and saw.

• Draw a labelled diagram to show how your periscope works.

• Write a set of instructions that explain clearly how to make your periscope.

When light meets a transparent object: refraction

Think like a scientist!

When light meets a transparent object it is not reflected. It travels through the object.

The speed of the light ray changes at the **boundary** (edge) of two different materials. For example, it changes speed when the light ray goes from air into glass, or from water into air.

Depending on the materials the light travels through, different things can happen to the speed and direction of the light ray. Because the speed changes, this often means that the direction of the ray changes. We call this change in speed and direction **refraction**.

Refraction can make objects appear differently when we look at them.

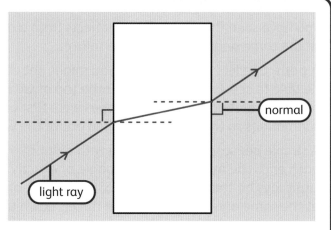

normal

light ray

Scientific words

boundary
refraction

1

Rank these materials in the order that you think light rays would travel through them, from the fastest to the slowest. Explain your reasons.

glass

water

diamond

acrylic (plastic)

air

Challenge yourself!

Do you think the speed of light in ice will be faster or slower than it is in liquid water? Explain your reasons.

See if you can find out the answer. Is this what you expected? Why do you think this is?

2

You are going to do a role play to explain what happens to a light ray when it is refracted.

> **You will need...**
> * piece of chalk

Work in a group.

a On the ground, draw a boundary line with your chalk.

b Label one side of the boundary line *air,* and the other side of the boundary line *glass.*

c Link arms as a group on the air side and face the boundary line so that you are standing parallel to it. See picture A.

d Walk forward together. What will you do when you pass from the air to the glass? Do it again, but this time walk from the glass to the air side. What will you do differently this time?

e Now stand, with your arms still linked together, on the air side of the boundary line. Stand at an angle to the boundary line. See picture B.

f Walk forward together. What happens when the first person meets the boundary line? Do the same thing again, but this time walk from the glass to the air side. What happens?

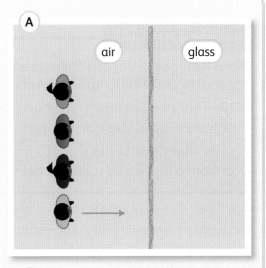

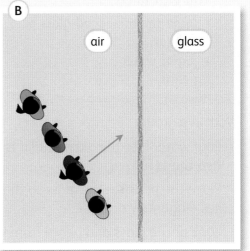

3

a Draw ray diagrams to represent what happened when you walked:

* straight from air to glass
* straight from glass to air
* at an angle from air to glass
* at an angle from glass to air.

b Some Class 6 learners have drawn ray diagrams to show what happens when light travels across the boundaries of different materials. Which diagrams are correct and which diagrams contain errors?

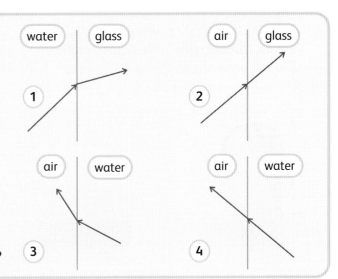

Investigating refraction

Think like a scientist!

When light goes from one transparent material to another, its speed can change:

- Light that travels from a solid to a liquid, or from a liquid to a gas, will get faster.
- Light that travels from a gas to a liquid, or from a liquid to a solid, will get slower.

1

You will need...
- large smooth-edged glass or jam jar
- jug of water
- pencil

Work in a group.

a Pour water into the glass until it is two-thirds full.

b Place the pencil in the glass so that it lies diagonally across it.

c Look at the glass and pencil from the side. What do you see? Make careful observations. Produce a drawing and notes of what you see.

d Think about what happens when light travels through one material to another. Explain your observations.

Let's talk

Discuss these questions with a partner:

a Have you ever seen how objects in or behind transparent materials can look different to how they actually are? How do they look different? Use scientific vocabulary to describe this.

b Through which different types of transparent material was the light travelling when you saw the objects looking different?

2

You will need...
- large smooth-edged glass or jam jar
- small smooth-edged glass (it needs to be able to fit inside the larger one)
- cooking oil

Work in a group.

a Pour cooking oil into the large glass, to a level that is lower than the top of the smaller glass.

b Stand the smaller empty glass in the cooking oil in the larger glass. Look at the smaller glass through the front of the larger glass. What do you see?

c Keep looking through the front and slowly pour cooking oil into the empty small glass.

d What do you see? Make careful observations. Produce a drawing and notes of what you see.

Science in context

Bioluminescence

People are not able to create their own light. We have to rely on light sources, such as the Sun and light bulbs, to create light for us. However, there are many animals that can create light by themselves. Scientists study these animals to see what we can learn from them to help humans develop better ways of lighting up the world.

The ability of an animal to create its own light is known as **bioluminescence**. A well-known bioluminescent animal is the firefly.

Scientists have also found that under the right conditions bubbles can give off light. Bubbles do this through a process known as **sonoluminescence**. Bubbles are little pockets of gas. When sound waves pass through them, they squash the bubbles. The bubbles then release energy in a fantastic burst of heat and light. This happens in nature when snapping shrimps clamp their claws shut.

The findings of these scientists might yield cleaner and more efficient sources of energy for us in the future. This would save fuel, and help to protect our planet from climate change.

firefly

snapping shrimp

Scientific words

bioluminescence
sonoluminescence
luminescent

Let's talk

a Discuss the words *bioluminescence* and *sonoluminescence* with a partner.

b What do you think *bio-*, *sono-* and *luminescence* mean?

c Discuss your ideas with another group.

1

Scientists have been able to make materials that are **luminescent** (give off light).

You are going to design an item of luminescent clothing for someone to wear.

a Think about a specific person who might find it useful to wear an item of clothing that gives off light. Will it give off light all the time, or will they be able to turn it on and off when they need to?

b Design your item of clothing. Produce an annotated poster showing where the luminescent material is and how it helps the person for whom you have designed it.

Science in context

Light poverty

Over 1 billion people on the planet live in light poverty, without access to **artificial** light once the Sun has set.

Without artificial light, hospitals cannot treat people, children cannot study, shops cannot do business, and we cannot move safely about at night. It is estimated that living without artificial light results in 1.5 million deaths every year.

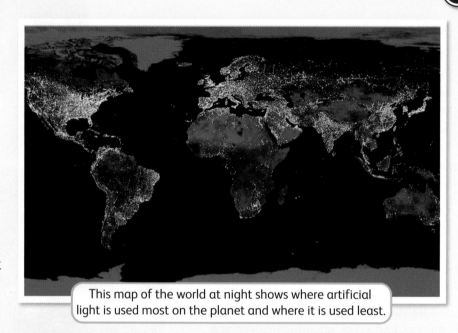

This map of the world at night shows where artificial light is used most on the planet and where it is used least.

The Sun is a free source of energy so scientists are working to develop solar-powered lighting solutions. Scientists are working on ideas for reducing light poverty that combine **solar panels**, **hydroelectric power** and **LED lighting**. LED lighting is cheaper, more efficient and healthier for humans and the planet than older types of light bulbs because it provides the best lighting using the least energy.

Scientific words
artificial
hydroelectric power
LED lighting
solar panels

1

You are going to research the energy-saving lighting solutions scientists are developing to reduce light poverty.

Work in a group of three.

a   Each of you must choose one of these options: solar panels, hydroelectric power and LED lighting.

b   Do some research to find out how your chosen technology can help reduce light poverty.

c   Make a flower with a different fact on each petal.

d   Share your findings with the rest of the group.

What type of scientific enquiry is Activity 1? Discuss what features of the enquiry make you think this with your group.

What have you learnt about light, reflection and refraction?

1

Think further about how you use artificial light.

a How many different things do you do after dark for which you need artificial light?

b How difficult would it be to do these things if you did not have access to artificial light in your home, community or city?

c Why do you think it is difficult to build power stations and electric grids in many countries in the world?

2

a Draw a double bubble map to compare and contrast the similarities and differences between reflection and refraction.

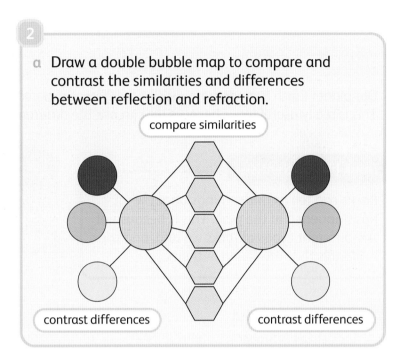

compare similarities

contrast differences   contrast differences

3

a What do you know now about reflection and refraction that you did not know before?

b What is the most interesting thing you have learnt about reflection and refraction?

4

Do some research to find out about:

a everyday devices that use reflection

b everyday devices that use refraction.

What can you do?

You have learnt about light, reflection and refraction. You can:

✔ state what happens when light meets an opaque object.

✔ describe how a ray of light changes direction when it is reflected from a mirror.

✔ state what refraction is and when it happens.

✔ describe how the direction of a light ray will change when it travels through different transparent materials.

✔ explain why the direction of a ray changes.

✔ describe what light poverty is, and how scientists are working to develop solar-powered lighting solutions.

The properties of rocks

What do you remember about rocks?

Our planet Earth is made from rock, and different rocks can have different properties. This is really useful for us because we can use the different rocks in different ways.

Think like a scientist!

Rocks can tell us many things about what has happened in the past. This is because of the different types of rocks we find, and the materials that are found in them.

Let's talk

a Take it in turns with a partner to name as many different rocks as you can.

b Write your list of rocks, and compare them with another group's list.

1 **You will need...**
- strips of paper
- coloured pens or pencils
- sticky tape

With a partner, you are going to make a paper chain showing what you can remember about rocks.

a What words can you remember that are linked to the properties of rocks?

b Make a long list – see if you can remember more than 10 words.

c Now write each word on a strip of paper – use one strip of paper for each of your words.

d Draw and label an image or write some facts about the word on your strip of paper.

e Now look at the words, organise them by finding words that link, for example *rock* and *hard*.

f Every time you add a new link, it must link to the word before it.

g Once all your links have been put together to make your rocks paper chain, share it with another group.

h Compare your chains, and add any words that they have which you do not have, to your own chain.

Earth: Our rocky planet

Think like a scientist!

The Earth is made up of four different layers. These layers are called the **crust**, the **mantle**, the **outer core**, and the **inner core**.

The Earth's crust is made from much cooler rocks than the mantle. Although the rocks at the surface can feel cool or cold, miners and cavers can feel an increase in temperature as they go down into the Earth's crust.

The mantle is very hot and thick, and is made of rocky material and **magma**. Magma is made of **molten** (melted) rock. The material in the mantle behaves like a

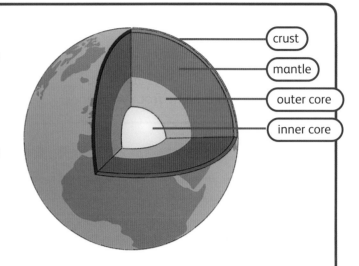

crust
mantle
outer core
inner core

very thick liquid that flows, a little like toothpaste does when you squeeze the tube gently. When this molten rock (magma) escapes to the Earth's surface, it is called **lava**.

Scientists have not yet been able to drill all the way through the Earth's crust.

The reason scientists know so much about the Earth is because of a number of models they use to explain the different evidence that they collect.

Models help us to understand and describe scientific phenomena and ideas.

This cave is in Sichuan Province, China. As the caver goes deeper, it becomes warmer and warmer.

Scientific words

crust
mantle
outer core
inner core
magma
molten
lava

Let's talk

We can think of a model of the Earth like a wrapped chocolate sweet that has a liquid centre with a nut in the middle.

Discuss these questions with your partner:

a How does the sweet represent the Earth?

b What are the good points about using this sweet as a model?

c What are bad points about using this sweet as a model?

Rocks from volcanoes

Think like a scientist!

Mount Etna is a volcano that still erupts. It is still throwing out molten rocks.

Obsidian, basalt and pumice are three different rocks that come from volcanoes.

1

You will need…

- a chunk of candle wax (red looks the best, but any colour will do)
- sand
- jug of cold water
- heatproof clear glass container (a mug, jug or beaker may work well)
- hotplate/Bunsen burner as a heat source

You are going to make a model to represent an underwater volcano. Your model will help you observe how molten rock can escape through an opening in the Earth's crust.

a Place the wax at the bottom of the heatproof glass container.

b Cover the wax with enough sand so that it cannot be seen.

c Carefully pour in water until the beaker is about three-quarters full.

d Place the beaker on the cold hotplate. Turn it on to medium heat.

e Watch carefully and make a record of how the contents of the beaker change over time.

f Once your volcano has stopped erupting, turn off the hotplate. Leave the glass beaker to cool completely before removing it from the hotplate.

Work safely!

Ask an adult for help with hot objects.

Place the glass on the hotplate before turning the hotplate on. Heat the glass slowly on medium heat, otherwise it may crack and break.

2

a Draw a cartoon strip to show what happened as you heated your volcano.

b Label each of the different parts of your model and describe what they represent.

c Describe how the crust is different from the mantle.

d Describe how magma is different from lava.

e Explain how magma is different from lava.

f Do you think this is a good model for an underwater volcano? Why?

Different types of rocks: Igneous rocks

Think like a scientist!

There are three types of rocks on Earth: **igneous** rocks, **sedimentary** rocks and **metamorphic** rocks.

These rocks are formed in different ways. Each type is able to change into another type.

Igneous rocks form when magma from a volcano cools and turns into a solid. This process is called **solidification**. As the hot rock cools down to form igneous rocks, crystals form inside the rocks. There are two kinds of igneous rocks:

basalt   granite

- extrusive igneous rock, which form outside the Earth's crust, like basalt
- intrusive igneous rock, which form within the Earth's crust, like granite.

Scientific words

igneous	sedimentary
metamorphic	solidification

1

You will need…
- blocks of dark, milk and white chocolate
- source of hot water, such as a kettle
- three pieces of aluminium foil, or three aluminium foil pie cases
- plastic bowl to hold the hot water
- timer or stopwatch

You are going to model how to make igneous rocks by using chocolate.
Once you have made your igneous rocks, keep them safe to use again later.

a Break up one type of chocolate into chunks and put the chunks inside a piece of aluminium foil. Now do the same for the other types of chocolate you are using.

b Heat the water and, once it is very hot, pour it carefully into the plastic bowl. If the kettle is heavy, ask your teacher to do this for you.

c Carefully place each aluminium foil case, containing chocolate, into the bowl, so that each one floats on top of the water.

d Observe and time how long it takes for the water to heat and melt the chocolate until a smooth liquid forms. Do all of the chocolates melt at the same rate?

e Carefully remove the aluminium foil with your molten chocolate from the bowl and leave it to cool. Your melted and cooled chocolate is now igneous rock. Keep your chocolate igneous rock for later.

f Is this a good model of how igneous rocks are formed? Explain the reasons for your answer.

Work safely! ⚠️

Take care when working with hot water!

Sedimentary rocks

Think like a scientist!

Sedimentary rocks are formed from **particles** (small pieces) of materials over thousands or even millions of years.

Rocks break down over time into **sediments** (small rock particles), and **erosion** (the movement of broken-down rock from one place to another) then happens. Erosion happens by wind, water and ice, moving the sediment to rivers, lakes or oceans.

The sediment then settles at the bottom of the water, a process known as **sedimentation**. Over time, all of the sediment creates layers, like layers of sand.

Burial of the sediment then happens as the thickness of the layers increases over time.

Over millions of years, the particles of rock become squashed together and form firm rock.

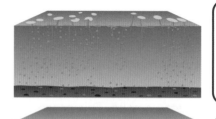

Sedimentation happens when sediment (small particles of rock) settle down in the water.

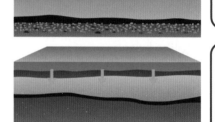

Burial happens as sediment layers build up over time.

Over millions of years, the particles become squashed together and form firm rock.

1

You will need...

- the three chocolate igneous rocks you made earlier
- cheese grater
- aluminium foil or aluminium foil pie dish
- metal spoon

Work safely! ⚠️

Take care when using the cheese grater.

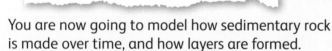

You are now going to model how sedimentary rock is made over time, and how layers are formed.

a Carefully remove your three chocolate igneous rocks from the aluminium foil pie dishes and grate half of each of them.

b In your new aluminium foil pie dish, build up different layers of the different colours of grated chocolate.

c When you have built up layers to fill the pie dish, carefully press down on the top of the chocolate with your spoon. Your layered and squashed chocolate is now sedimentary rock. Keep your chocolate sedimentary rock in the aluminium foil pie dish. Also keep what is left of your igneous chocolate rocks; you will use them later.

Scientific words

particles
erosion
sediments
sedimentation
burial

Model erosion and sedimentation

1

You will need...
- piece of guttering with two end pieces
- wooden block or something similar to raise one end of the guttering
- sand and gravel, or sand and soil mixture
- plastic tank
- jug of water
- timer or stopwatch
- hand lens

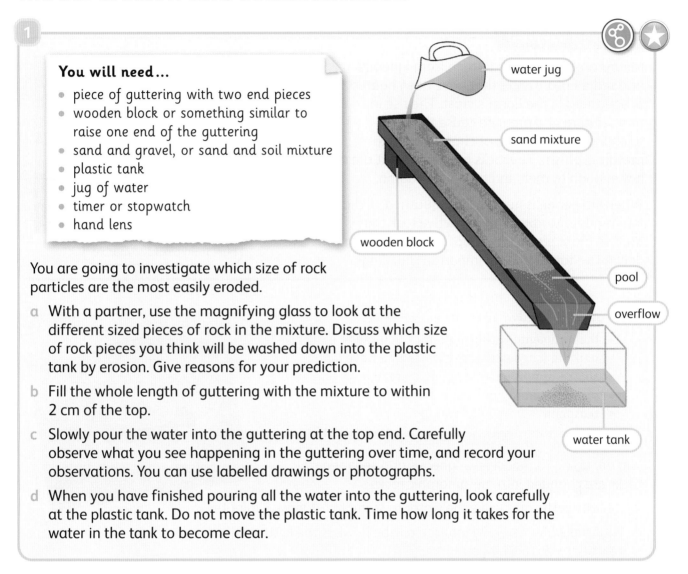

water jug

sand mixture

wooden block

pool

overflow

water tank

You are going to investigate which size of rock particles are the most easily eroded.

a With a partner, use the magnifying glass to look at the different sized pieces of rock in the mixture. Discuss which size of rock pieces you think will be washed down into the plastic tank by erosion. Give reasons for your prediction.

b Fill the whole length of guttering with the mixture to within 2 cm of the top.

c Slowly pour the water into the guttering at the top end. Carefully observe what you see happening in the guttering over time, and record your observations. You can use labelled drawings or photographs.

d When you have finished pouring all the water into the guttering, look carefully at the plastic tank. Do not move the plastic tank. Time how long it takes for the water in the tank to become clear.

2

Using your results from Activity 1, discuss these questions with a partner:

a Was the water in the plastic tank clear or cloudy when you had finished pouring all the water? Why do you think this is?

b What happened to the water in the plastic tank over time? Why do you think this happened?

c Was your prediction correct? Which sizes of rock pieces were moved into the plastic tank? Why do you think this happened?

Let's talk

Discuss steps **a** to **c** in Activity 1 with a partner. Explain what each step of the model is trying to demonstrate about sedimentary rocks.

Metamorphic rocks

Think like a scientist!

Metamorphic rocks are formed from igneous and sedimentary rocks that have been heated or squashed in the Earth's crust. The rise in temperature and pressure cause the rocks to change. This process of change is called **metamorphism**. The rocks are heated but are not hot enough to melt and turn into magma.

When limestone is heated and squashed, it changes into marble. Marble is an attractive rock that is used to make statues, the tops of expensive and decorative tables and for important buildings.

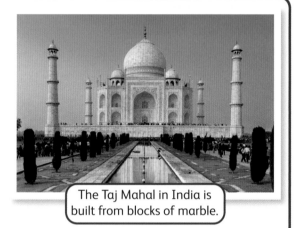

The Taj Mahal in India is built from blocks of marble.

1

You will need...

- the sedimentary chocolate rock you made earlier and the igneous chocolate rock you have left
- source of hot water, such as a kettle
- plastic bowl to hold the hot water
- teaspoon

You are going to model how igneous and sedimentary rocks are changed into metamorphic rocks.

a Boil the kettle and leave the water to stand for at least 5 minutes.

b Pour the water carefully into the plastic bowl. If the kettle is heavy, ask your teacher to do this for you.

c Add the igneous chocolate rock you have left to the top of the sedimentary chocolate rock in the aluminium foil pie dish. Carefully place the pie dish so that it floats on top of the hot water.

d Observe as the water heats the chocolate and the chocolate starts to melt.

e Carefully remove the aluminium foil pie dish from the bowl when the chocolate is soft to the touch. Use your spoon to gently test for when this happens.

f Before the chocolate cools down, gently squash and fold the mixture with your spoon. Your heated and squashed chocolate is now metamorphic rock.

g Is this a good model of how metamorphic rocks are formed? Explain the reasons for your answer.

Scientific word
metamorphism

Work safely! ⚠

Take care when working with hot water!

Fossils in rocks

Think like a scientist!

Some rocks contain amazing things, such as **fossils**. Fossils are the dead remains of parts of plants and animals from millions of years ago. The plant or animal was covered quickly after death by sediments such as mud and sand. In time, the sediments hardened to form rock and the bodies inside the sediments formed fossils, called **cast fossils**.

fossil of a fish

Some fossils are not formed by the actual plant or animal body but by the things that they left behind, such as eggs, solid wastes (their droppings) and tracks. These fossils are known as **trace fossils**.

1

You are going to work out how fossils have been formed over millions of years.

Work in a small group.

These mixed-up pictures and sentences describe how fossils form.

a Use the internet or books to research how fossils form.

b Match the sentences below (numbered 1 to 4) to the pictures and put them in the correct order.

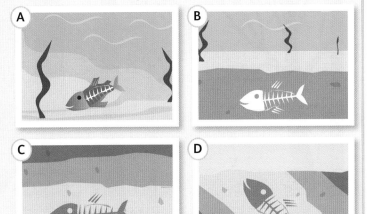

1 The hard parts of the animals remained. They were squashed and covered by new rock.

2 More bits of rock that were washed down by rivers covered the rotting animals.

3 After death, mud or sand covered some animals.

4 After a very long time, the land changed and the animal remains changed to rock. We call these fossils.

Let's talk

Fossils are generally only found in one of the types of rocks.

Discuss with a partner which type of rock you think it is and why. Use what you have learnt about the three different types of rock.

Scientific words

fossils
cast fossils
trace fossils

The fossil record (1)

Think like a scientist!

Geologists are scientists who study rocks and soil. **Palaeontologists** are scientists who study fossils found in rocks. They have discovered that some fossils are only found in one or a few layers of rock, while a few were found in many layers of rock.

Fossils found in one layer meant that the organism which formed them lived in just one time period. Palaeontologists also found some of these fossils in many parts of the world, and they are easy to identify. This makes these fossils useful in helping to indicate the age of the layer of rock in which they were found, so these types of fossils are called **index fossils**.

Each group of living things has a **fossil record**. This shows when members of the group existed and left fossils behind. Some groups of living things, like algae and bacteria, have a long fossil record. The mammal group has a much shorter fossil record.

The fossil record does not give an exact time of when scientists think the Earth formed, but it does suggest that the Earth is very old.

The artist drew this picture after studying fossils from the Carboniferous period and imagining how the living things may have looked in their environment.

Let's talk

Look at the picture above and discuss these questions with your partner:

a   What types of plants grew during this period?

b   What types of animals were found then?

c   What do you think the climate was like then? Explain why you think so.

Now find out more about index fossils, and about different time periods in the fossil record, on the next two pages.

Scientific words

geologists   palaeontologists
index fossils   fossil record

The fossil record (2)

Think like a scientist!

The pictures below show the index fossils for the different time periods into which scientists divide the fossil record. It shows the general name of the group of animals to which each fossil belongs.

Time period	Index fossil
Quaternary	scallop
Tertiary	sea snail
Cretaceous	ammonite
Jurassic	ammonite
Triassic	ammonite

Time period	Index fossil
Permian	brachiopod
Carboniferous	coral
Devonian	brachiopod
Silurian	coral
Ordovician	trilobite
Cambrian	trilobite

Challenge yourself!

Work in a group.

a Do some research, using a dictionary or the internet, to find out how to pronounce the names of the different time periods correctly.

b Challenge another group to a mini-quiz to see which group can pronounce the most names correctly.

1

a Which time period is identified by a scallop?

b What does this tell you about what the area where the scallop fossils were found was like long ago?

c Which time periods have ammonites as their index fossils?

d If you found a rock layer with a brachiopod in it, what index fossil would you look for in the layer beneath?

113

The fossil record (3)

Think like a scientist!

This bar chart shows the fossil record of living things as they appeared and disappeared from the Earth during the different time periods.

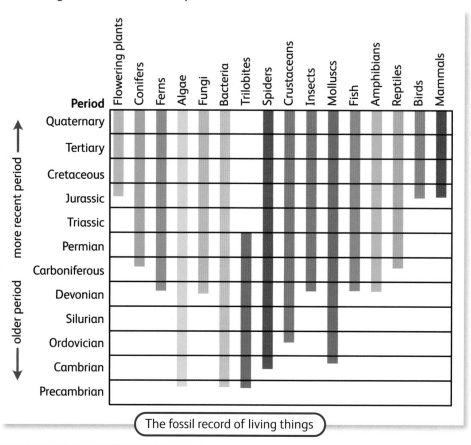

The fossil record of living things

Let's talk

Use the fossil record chart above to discuss these questions in groups:

a Which appeared in the fossil record first – ferns or conifers?

b How many new groups were added to the fossil record in the Devonian Period?

c Which group has become extinct, and when was it last recorded in the fossil record?

d In the period during which the flowering plants appeared, which animal groups also appeared in the fossil record? Why do you think this is?

fossil of a plant

Challenge yourself!

Try and find out what Pangaea is, and how fossils provide evidence that it existed!

How our scientific understanding of fossils developed over time

What we know about science today is different to what people thought many years ago. Over the years, scientists have found out more and more about the world and realised that what people once thought was not always true. So, our understanding of science has changed over the years.

This fossil is an ammonite.

Ammonites were first thought to be the curled-up bodies of stone snakes. Later they were found to be the shells of creatures that swam in ancient seas. Only the hard part of the animal survived and turned into a fossil.

The word *fossil* was first used by Georgius Agricola (1494–1555), a German doctor. He used the word to describe anything that was dug out of the ground, including ancient pottery. Today the word fossil means the remains or trace of a prehistoric plant or animal embedded in rock.

Fossil of a shark tooth

Long ago, people found triangular-shaped stones, which they called tongue stones. Some people believed these stones grew in the rocks and others believed they fell from the Moon. Nicolaus Steno (1638–1686), a Danish geologist, noticed that these stones were similar to the teeth of a shark. This observation led him to believe that the tongue stones were the teeth of ancient sharks, left behind in the rock after the shark had died. He then looked at other fossils and decided that they too had come from the bodies of ancient animals.

Let's talk

Discuss these questions in a group:

a What scientific enquiry skills did Nicolaus Steno use?

b How did the ideas Steno proposed help us develop our understanding of the past?

c Why do you think most scientists in Mary Anning's time were men?

d Do women today have the same opportunities as men to become scientists? Explain why you think so.

Mary Anning

The first known fossil collector was Mary Anning, who lived in England from 1799 to 1847. She discovered many fossils of sea dinosaurs and sold them to museums. At a time when British scientists were mainly rich men from London, this woman from a small town became one of the best fossil hunters ever.

$\rightarrow$

Living fossils – the coelacanth

Some plants and animals still alive today are living fossils. They have remained mostly unchanged since the earliest prehistoric times.

In 1938, 32-year-old Marjorie Courtenay-Latimer worked at a tiny museum in East London, South Africa. She noticed an unusual and beautiful blue fish among the day's catch that a fisherman showed her. From her reference books, she thought it looked similar to a prehistoric fish only known from fossils. She posted a rough sketch and description of the fish to Professor J.L.B. Smith at the university in Grahamstown, South Africa.

Prof. Smith identified that the fish was indeed a coelacanth, which everyone thought had died out long ago. He realised

Fossil of a coelacanth

that there must still be living coelacanths. Since then, fishermen have found more than 50 living coelacanths!

New fossil discoveries continue to improve and even change our understanding of the fossil record and the development of living things on Earth, like these two discoveries by children!

- Matthew Berger was just 9 years old when he tripped over fossil remains while chasing his dog in an area where his father often hunted for fossils. The bones belonged to a young boy and young woman who had died at the same time. Sediments then covered their remains deep inside the Malapa cave north of Johannesburg, South Africa, preserving their bones for nearly 2 million years.

Human fossil remains

- In 2004, 7-year-old Diego Suarez from Chile found some pieces of backbones from a prehistoric creature on a research trip with his geologist parents.

1

Work in groups.

a On a large, long sheet of paper (like brown wrapping paper), create a timeline of the different fossil discoveries discussed on pages 115 and 116, from the earliest to the most recent. Leave space to add some new discoveries.

b Do some research about other new fossil discoveries. Add these to your timeline.

Let's talk

Discuss these questions in a group:

a Do you think it was just good luck that helped Marjorie Courtenay-Latimer, Matthew Berger and Diego Suarez to make their fossil discoveries? Why or why not?

b What can you learn from the timelines you created in Activity 1 about how our scientific understanding of the fossil record has changed over time?

The rock cycle

Think like a scientist!

Where do the rocks in the Earth's crust come from? How do the different types form? These questions puzzled scientists for a long time. As they looked at the different types of rocks and where they were found, they began to build up an answer.

Scientists realised that rocks of one kind can change into rocks of another kind. They called this the rock **cycle**.

Let's talk

In science, we use cycles to help us explain a regularly repeated sequence of events. Diagrams of cycles have arrows in them, such as the life cycle of a butterfly.

a What other cycles can you think of that you have heard about in science?

b What do the arrows in the cycles you can think of show?

1

In pairs, you are going to be rock detectives. You will apply all your learning about rocks, and like the scientists, work out the rock cycle.

a As a pair, you are going to produce a poster showing the rock cycle and how rocks change from one type into another (remember to use arrows).

b One of you will be the illustrator and produce labelled drawings. The other one will write the text, using the key words to explain what is happening. The words you need to include in your poster are:

igneous   magma   metamorphic   volcano   sedimentary   melting

solidification   extrusive   intrusive   sedimentation   erosion

metamorphism   burial   water   wind   ice

c Compare your poster with another pair's poster. Are your ideas the same or different? Is there anything you could add to your poster?

Scientific word
cycle

2

You will need...
- chunks from different chocolate
- paper plate

Use models to represent and apply your ideas about the three different types of rocks that we find on Earth.

a In your group, examine the different chunks of chocolate.

b Sort the chunks into igneous, sedimentary and metamorphic chunks on the paper plate. Explain your reasons for sorting them as you have.

Rocks make soils

Think like a scientist!

Rocks are broken down by rain, wind and water by a process called **weathering**. Weathering of rocks can form **soil**.

In Class 1, you learnt that soil is a mixture of tiny ground-down pieces of rock, dead plants, water and air.

Soil can be black, red, yellow, green, brown and even purple! The type and colour of soil depends on the rocks that made it.

Soil is usually covered by plants that are growing in it. Soil allows plants to set down their roots and hold their position in a habitat. It also stores water and minerals for the plants to use as they grow. Successful farming depends on the soil, so over the years, scientists have studied soil to see how it can best be used to grow crops.

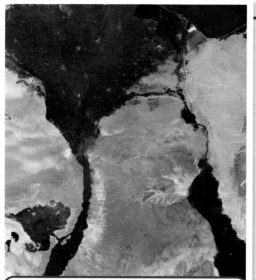

The rocky fragments carried down the river Nile have settled out and formed a delta where the river meets the Mediterranean Sea. A rich soil has formed which has been farmed for over 5 000 years.

This farmer in northern Peru is growing yams in a deep, crumbly, well-drained soil. These soil properties are ideal for his crop.

1 **You will need...**
- small bags or cups
- pen
- spoon
- hand lens

Work safely! ⚠

Always wash your hands after handling soil. Put any animals back where they came from.

Scientific words
weathering
soil

You are going to be a soil scientist!

a Walk around your school grounds. Look for different types of soil.

b Use a spoon to take small samples of different kinds of soil. Put each different sample in a small bag or cup, and label where you got each sample from.

c Talk to your group. Look at different samples from different places. Use a hand lens.

d What is the same about them, and what is different? Think about colour and texture. Are there stones or parts of plants in your soil samples?

e Are there any animals in your soil samples? What kind of animals? Why do you think they are in the soil?

2

You will need...

- different coloured soil samples dried in air
- pestle and mortar
- spoon
- plastic or paper cups
- fine mesh such as tights
- elastic band
- PVA glue
- water
- paintbrush, sponge and/or rags
- paper

Cave painting from the Matobo Hills in Zimbabwe

Soil colours serve as dyes in bricks, pottery and artwork. You are going to create a painting using soil.

a   Add one of the dried samples of soil to the pestle and mortar and grind it up. The finer it is ground up the better.

b   Wrap the top of a cup in the fine mesh and secure it with an elastic band.

c   Carefully spoon your ground-up soil onto the top of the cup. Shake it gently so the finest powder passes through. Any lumps that are too big to pass through can be thrown away.

d   Remove the mesh. Add some PVA glue to your fine soil and mix it together. If it is too thick, add a small amount of water.

e   Repeat steps **a** to **d** for all the different coloured soils you have.

f   You are now ready to paint your picture using your soil paints!

Ancient rock painting from the Cave of the Hands in Argentina

Challenge yourself!

Red ochre is one of the oldest pigments (natural colours) still in use. It was first used as an artistic material, as far as we know, in prehistoric cave paintings.

Try to find out in what types of soil red ochre can be found, and what gives it its red colour.

Create your own picture with your soil paints!

Soil layers

Think like a scientist!

Soil is divided into different layers.

At the surface is the darkest layer. It is made mainly from **organic** materials (the litter or remains of dead plants and animals). Bacteria in the soil feed on these remains and break them down. As the remains of plants and animals rot, they form a substance called **humus**.

The thicker layer under the surface is a mixture of humus and rocky fragments and is called the **topsoil**. It is the part of the soil that farmers and gardeners use for growing their plants. It is also the layer that ecologists examine when they are studying a habitat.

Below the topsoil is the **subsoil** (*sub* means *under* or *below*). This layer is paler than the topsoil because it contains much less humus. Below the subsoil is a layer containing lumps of rock. This layer is called **bedrock**.

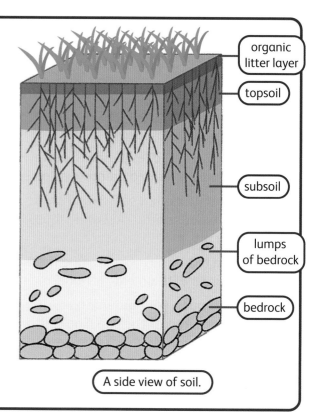

organic litter layer

topsoil

subsoil

lumps of bedrock

bedrock

A side view of soil.

1

You will need...
- spade
- pencil and paper
- ruler

You are going to see if you can identify the different layers of soil in your school grounds.

a   In a group, dig a hole deep enough for you to see the first three layers of soil – the thin layer of dead plants and animals, the topsoil and the subsoil.

b   Measure the depth of each layer of soil.

c   Make detailed observations of the three layers and record them in a drawing. Include information about the colours. Can you see anything in the layers? What is the texture of each layer like?

d   Fill the hole back in and go and do the same in different places around the school grounds.

Work safely! ⚠

Remember to always wash your hands after handling soil.

Scientific words

organic	humus
topsoil	subsoil
bedrock	

Different types of soil

Think like a scientist!

Sand, silt and clay are **inorganic** materials that do not come from living things.

They are made up of tiny rock particles:

- **Sand** is made up of larger particles that we can see with the naked eye. Sandy soils feel gritty.

- **Silt** is made up of particles too small to see with the naked eye. Silty soils are often found in places that have flooded and dried out again. Silty soils feel silky.

- **Clay** is made up of tiny particles that fit together very closely. Clay soils feel sticky. Clay soils are not as easy to squeeze together as the other types of soil.

Many soils are made up of a mixture of different soil types. **Loam** has a mixture of 40 % sand, 40 % silt and 20 % clay. It also has large amounts of organic materials, called humus. These organic materials come from the remains of dead plants and animals. The humus binds the soil particles together, making loam soil crumbly. This gives loam soil special properties, making it the ideal soil for gardening and farming:

- Roots can grow more easily through this crumbly soil than through hard rocky soil or dense clay soil.

- The humus soaks up water like a sponge, so that plant roots can slowly absorb the water and the minerals dissolved in it.

- The mixture of sand and clay means that loam soil lets water drain away better than clay, which is important in wet weather. But loam soil can also hold some water better than sand when there is little rain.

loam soil

Scientific words

inorganic   sand   silt   clay   loam

1

Work in a group.

a Collect soil samples from different places in your community.

b Place each soil sample on a numbered sheet of paper.

c Carefully look at each soil sample:
  - Rub the soil between your fingers.
  - Smell the soil.
  - Look at the size of the soil particles with a magnifying glass.

d Write an identification card for each soil sample to explain what type of soil you think it is (sand, silt, clay or loam).

e Discuss why you think loam soil is the ideal soil for farming.

f Which of your soil samples would be good for farming? Explain why.

g Keep your soil samples to use for the activity on the next page.

Drainage in different types of soil

Think like a scientist!

The way water travels through soil is called **drainage**. The drainage of water is different through sandy, silty, clay and loam soils. You are going to investigate which soil allows water to drain quickly and which soil holds the water in it.

Choose some of your group's soil samples from Activity 1 on the previous page to use for this activity.

Scientific word
drainage

1

You will need...
- 2-litre plastic bottle
- scissors
- 8 coffee filter papers
- jug of water
- timer or stopwatch
- 4 cups of different soils – one sandy, one silty, one clay, one loam with organic matter

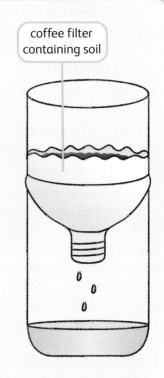

coffee filter containing soil

Work in a group.

a You are going to use sandy soil, silty soil, clay soil and soil containing organic matter. Predict which soil you think will let water flow through it quickest and which one will let water flow through the slowest. Explain your answer.

b How will you make the experiment a fair test?

c Make your filter, using the bottle.

d Put one of the coffee filters into the funnel. Add one of the soils, and place another filter over the top. This is to keep the soil from splashing and help the water to cover all the soil.

e Measure how much water you pour onto the soil sample (so you can use the same for the other samples).

Time how long it takes for the water to pass through.

f Make a table to record your results.

g Clean the funnel and insert a new filter.

Repeat the steps above with the different soils.

Was your prediction right or wrong? Why do you think this is?

Let's talk

Discuss the following questions with your group:

a Which of the soils that you tested would be best for growing crops on farms? Why?

b Which of the soils would cause problems with flooding during the rainy season if houses are built on it? Why?

Farming and soil

Soil is one of the Earth's most important natural resources.

Increased demand for food has led to forests and grasslands being changed to farmed fields and pastures.

Sometimes, when people have changed land from forests and grasslands to farmed land, the soil can be damaged and lost. This can happen if trees are cut down because the roots that held the soil in place are no longer there and the soil is washed away. Many of the new plants that are farmed, such as coffee, cotton, palm oil, soybeans and wheat, can actually increase this damage, called **soil erosion**.

Farming can damage the soil.

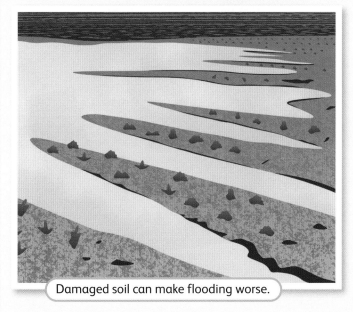

Damaged soil can make flooding worse.

In the last 150 years, half of the topsoil on the planet has moved from where it was originally. The land has therefore changed, plants can no longer grow, animals lose their habitats, and this can lead to new deserts being created.

As well as topsoil movement, erosion has also resulted in increased pollution and sedimentation in streams and rivers. This can lead to clogged waterways and a decline in fish and other species.

Eroded land is often less able to hold onto water, which can worsen flooding.

Soil quality is also being affected by farming. Pesticides and other chemicals used on crops have helped farmers increase yields (the amount of food they grow). However, scientists have found that the overuse of some chemicals changes soil composition and disrupts the balance of microorganisms in the soil. This stimulates the growth of harmful bacteria, and can hinder plant growth.

Scientific word
soil erosion

1 **You will need...**

- three large plastic bottles
- scissors
- three smaller plastic bottles with lids
- string
- soil

- twigs, bark, leaves, dead roots
- seedlings (geraniums, cress, basil or chives)
- spoon
- cups of water
- ruler

In a small group, you are going to investigate how forests help maintain soil quality and prevent pollution from getting into lakes, rivers and groundwater.

Work safely!

Be careful with scissors; ask the teacher to help if it is difficult to cut the bottle.

a Prepare three identical bottles by resting them on their sides on a flat surface and cutting off the side of each bottle, as shown in the picture. It may help to put an object to the left and right of each bottle to stop them from rolling.

b Fill each bottle with the same amount of soil. Press the soil down firmly with your spoon. The soil must be below the level of the bottle opening at the top of the bottle.

c Cut the tops off the three smaller bottles and carefully put holes in each side. Tie string to the three smaller bottles and hang them from the ends of the larger bottles, as shown in the picture.

d Plant your seedlings in the first bottle. Add twigs, bark, leaves and dead roots to the second bottle. Leave the last bottle with just the soil.

e Pour two cups of water into all three bottles.

f Record your observations every two minutes for each bottle. Make a note of how much water drains out and the colour of the water.

g You can repeat step **f** over several days and add more notes to your records.

h Using your results, what conclusions can you make about how forests affect the quality of water that runs out of them?

i Produce a video report to explain the importance of vegetation for soil and water.

What have you learnt about rocks and soils?

1

In small groups, play a game about rocks.

a   Your teacher will put up posters showing the three different types of rocks and magma.

b   In your group, come up with as many questions as you can where the answer could be *igneous*, *sedimentary*, *metamorphic* or *magma*. Try and make up some tricky questions. Hand all your questions to your teacher.

c   The first person from each group should go to the front of the class.

d   Your teacher will read out a question from one of the groups, and then say 'Go'. The person at the front of the class should go and stand next to the poster that represents the answer.

e   Every correct answer scores a point. It is not a race, so everyone standing by the correct answer gets a point.

f   The second person from each group should go to the front of the class and the teacher should read out the next question, and so on.

g   The group with the most points wins!

2

Write three paragraphs. Each paragraph needs to include five of the key words from the list below. Your paragraph should explain how the key words link to each other.

crust   mantle   outer core

magma   inner core   molten

fossils   igneous   sedimentary

metamorphic   solidification   soil

erosion   burial   sedimentation

lava   rock cycle   metamorphism

organic   humus   topsoil

Challenge yourself!

Produce a mind map that links all the key words in the box above. You may need a large sheet of paper!

What can you do?

You have learnt about rocks. You can:

✔   name the three different types of rocks.

✔   describe the features of the different types of rocks and explain how this makes them useful to us.

✔   describe in which type of rock fossils are formed, and how they are made.

✔   describe the rock cycle.

✔   explain how one type of rock can change into another.

✔   explain why water flows through different types of soil at different rates, and how different soils can support or hinder plant growth.

The Earth, Sun and Moon

What do you remember about the Earth, Sun and Moon?

Think of five things that you know about the Earth, Sun and Moon.

Think like a scientist!

In this unit, you will learn about our place in space. This includes our Solar System (our neighbourhood in space) and its place in the universe. You will find out about the different positions and movements of the planets, the Moon and the Sun, and how the Moon appears to change shape over time.

Let's talk

Look at the four diagrams. In a group, discuss which diagram correctly shows the Earth, Sun and Moon. Explain your thinking.

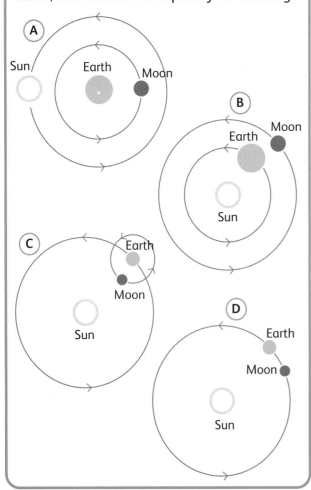

1. You are going to model the motion of different objects in the Solar System.

   a Work in a group of three. Find an open space.

   b Decide who will be the Sun, the Earth, and the Moon.

   c The 'Sun' stands still in the middle of the open space.

   d The 'Earth' walks around the Sun in an **anticlockwise** direction. While walking around the Sun, the 'Earth' also turns around anticlockwise.

   e The 'Moon' walks around the 'Earth' in an anticlockwise direction. While walking around the 'Earth', the 'Moon' turns around very slowly in an anticlockwise direction. The 'Moon' must always face the 'Earth'.

Scientific word

anticlockwise

Rotation of the Earth

Think like a scientist!

Planets are **rotating** (spinning) all the time. They spin around an imaginary line called an **axis**. The axis of the Earth is tilted at 23.5 degrees.

One complete **rotation** on the axis of the planet is called a **day**.

On Earth it takes 24 hours to make one complete rotation. This is the length of a day on Earth.

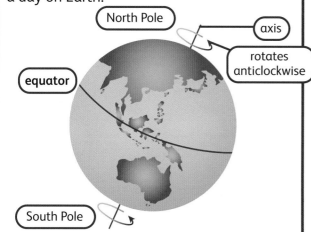

North Pole

axis

rotates anticlockwise

equator

South Pole

1

You will need...
- globe
- lamp
- two tables
- plastic figure
- sticky putty

You are going to model how daylight and darkness change during one day on Earth.

a Darken the room as much as possible. Place the globe on a table. Place the lamp on another table, in line with the globe. Switch on the lamp, making sure that the light shines directly towards the centre of the globe.

b Find the place where you live on the globe. Stick a plastic figure on the place where you live with sticky putty.

c Slowly rotate the globe anticlockwise. Observe the areas of light and shadow. Discuss the questions in the *Let's talk* box with your group.

Let's talk

Discuss these questions in a group:

a Which side of the globe represents daytime?

b What fraction of the Earth is in daylight/ darkness at any time?

c Name some countries where it is night time when it is daytime where you live. Use the globe and lamp to help you.

Scientific words

rotating	axis
rotation	day
equator	

Challenge yourself!

Imagine that you have a friend who lives on the other side of the world. What would be the best time of day to phone your friend? Make sure you choose a time when you are both awake!

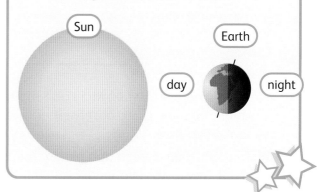

Sun

Earth

day

night

Our Solar System

Think like a scientist!

The **Solar System** is our neighbourhood in space. At its centre is a **star** (a huge ball of super-hot gas, that gives off light), the **Sun**. Objects that give off their own light, like the Sun, are called **luminous**.

Eight **planets** orbit the Sun. The Solar System formed about 4.6 billion years ago. It formed from a huge disc of gas and dust left behind after the Sun formed. Planets are rock or gas objects that **orbit** around a star, in fixed orbits. They are **spherical** in shape.

Near the Sun, where gravitational interactions were stronger, heavier elements clumped together to form the inner **rocky planets** – Mercury, Venus, Earth and Mars. Further out, the Sun's gravitational pull was weaker and lighter elements formed the outer planets – the **gas giants** – Jupiter, Saturn, Uranus and Neptune.

Venus, Mars and Jupiter can sometimes be seen with the naked eye from the Earth, looking like bright stars. However, they are not often all visible in the sky at once!

Scientific words
Solar System
star   Sun
luminous
planets   orbit
spherical
rocky planets
gas giants

Let's talk

Discuss these questions:

a Why do you think that Venus and the Moon appear brighter in the night sky than the stars beyond our Solar System, even though they are not luminous objects?

b Is there a way you could test your ideas?

1

Do you remember how a mnemonic can help you to remember something, like *Ripen Off Your Green Bananas In Vinegar* to remember the order of the colours in the rainbow (Red, Orange, Yellow, Green, Blue, Indigo and Violet)?

What mnemonic can you think of to remember the order of the planets from the Sun?

Challenge yourself!

a Find out where and when Mars and Jupiter will be visible in the night sky, and try to observe them. Keep a record of your observations.

b Pluto used to be classed as a planet, but is now a dwarf planet. Find out why it was changed. Explain it to someone else.

The Moon

Think like a scientist!

The Moon is a satellite of the Earth. This is because the Moon orbits around the Earth.

As the Moon orbits the Earth, it spins very slowly on its axis. It takes the Moon one full orbit of the Earth to also turn on its axis once. This is why the same side of the Moon always faces the Earth; it does not matter where on Earth you look at the Moon.

1

In a group, you are going to model how we always see the same side of the Moon, even though the Moon is spinning.

1 First, you are going to look at what we would see if the Moon did not spin.

2 One learner is the Moon. The rest of the group represent people on Earth, standing close together in the centre of a circle, facing outwards and all looking at the Moon.

3 The learner acting as the Moon needs to put both arms out to the side. This will help everyone see which way this learner is facing.

4 Look at the diagram opposite. The learner acting as the Moon should slowly (and carefully) walk in a circle around the Earth, stopping at positions A, B, C and D. The learner should not rotate while doing this, as shown in the diagram.

Discuss these questions in your group:

a What does everyone see when the Moon is at position A?

b What does everyone see when the Moon is at position B?

c What does everyone see when the Moon is at position C?

d What does everyone see when the Moon is at position D?

e Are you seeing the same or different sides of the Moon?

2

Now you are going to look at what we see when the Moon is spinning.

1 Again one learner is going to be the Moon, and the rest of the group will represent people on Earth, standing together in the centre of the circle, facing outwards and all looking at the Moon.

2 Again the learner acting as the Moon needs to put both arms out to the side.

3 Look at the diagram opposite. The learner acting as the Moon should slowly (and carefully) walk in a circle around the Earth, stopping at positions A, B, C and D. The learner should rotate slowly while doing this, as shown in the diagram.

Discuss these questions in your group:

a What does everyone see when the Moon is at position A?

b What does everyone see when the Moon is at position B?

c What does everyone see when the Moon is at position C?

d What does everyone see when the Moon is at position D?

e Are you seeing the same or different sides of the Moon?

f How many times has the Moon rotated when it has gone around the Earth once?

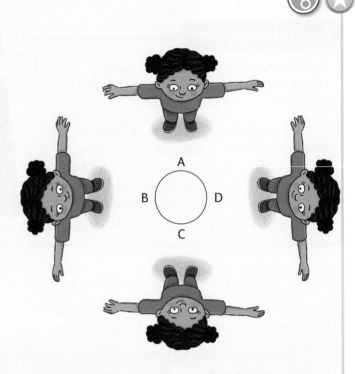

3

a Do some research to find out how long it takes the Moon to rotate once on its axis.

b To make this model better, what should the learners who are the Earth be doing while the Moon is going around the Earth once?

c Make a video, or 3D model, to explain why we can see the same side of the Moon from anywhere on the Earth, even though the Moon is spinning.

The Moon's orbit

Think like a scientist!

The Moon's orbit around the Earth is elliptical, so one side of the orbit is sometimes closer to the Earth than the other. As a result, the distance between the Moon and the Earth varies throughout the month and the year.

If we took a photograph from the same position on Earth every day, the Moon would appear to change in size and move to different places in the sky.

Challenge yourself!

Find out what we call the following:

- the point on the Moon's orbit closest to the Earth
- the point on the Moon's orbit furthest from the Earth
- what we call these types of Moon.

Challenge yourself!

The Full Moons that happen throughout the year have different names. The first Full Moon, in the northern hemisphere, in January is called the Full Wolf Moon. This is because ancient civilisations noticed that packs of wolves would howl at the Moon during the cold months.

Try to find out some more of the names for the Full Moons and the reasons why they were given these names.

Did you know?

The Moon's orbit is getting larger, at a rate of about 3.8 centimetres per year. The Earth's rotation is slowing down because of this. One hundred years from now, our day will be 2 milliseconds longer.

The first person to work out in a mathematical way how the Moon's orbit would change, was George Howard Darwin (Charles Darwin's son)!

The phases of the Moon

Think like a scientist!

When viewed from Earth, the Moon appears to change shape. These apparent changes in the shape of the Moon are called **phases**.

We can only see the sunlit side of the Moon. At different points in the Moon's orbit around the Earth, different amounts of the Moon's sunlit side face the Earth.

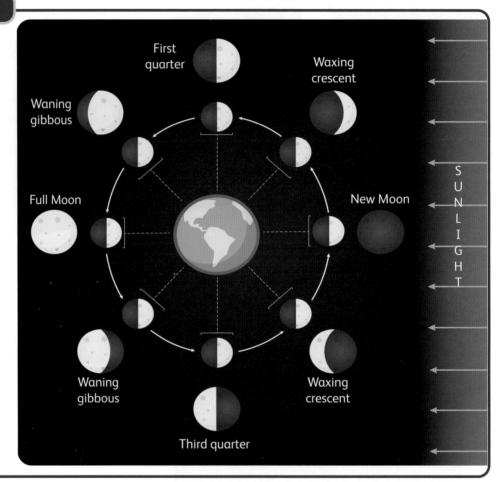

First quarter

Waxing crescent

Waning gibbous

Full Moon

New Moon

SUNLIGHT

Waning gibbous

Waxing crescent

Third quarter

1

You are now going to observe the shape of the real Moon.

a Observe the Moon on the next clear night. Record its apparent shape.

b Use the diagram in the *Think like a scientist!* box to work out where the Moon is in its orbit around the Earth.

c Predict how the Moon's apparent shape will change over the next few nights.

d Observe and record the apparent shape of the Moon for a few nights. Were your predictions correct?

e Continue observing and recording the apparent shape of the Moon until it is the same shape as it was when you started.

Scientific word
phases

2 **You will need...**
- dark room with a high up bright light (like a data projector)
- polystyrene ball on the end of a wooden stick

Scientific word
astronaut

You are going to find out why the shape of the Moon changes as it moves around the Earth.

a Hold the ball at arm's length so that it is in the light of the projector. The ball is your Moon and the light from the projector is your Sun. You are now looking up at the Moon from Earth.

b Hold the wooden stick so you keep the Moon in the same place and slowly walk around the Sun. What do you see? What parts of the model worked well in helping explain the Sun? What parts did not work well to explain the Sun? Why?

c Draw what the Moon looked like when you were in the four positions in the pictures opposite. If you cannot remember what the Moon looked like, go and look again at your Moon in the Sun's light.

d Which parts of the model worked well in helping to explain the phases of the Moon?

e What parts did not work well to explain the phases of the Moon? Why?

Did you know?

In 2019, NASA celebrated the 50th anniversary of the first landing on the Moon by a human. **Astronaut** Neil Armstrong stepped from the lunar module Eagle that had detached from their orbiting Apollo 11 spacecraft to become the first man on the Moon, followed by Buzz Aldrin. Michael Collins remained in orbit in the Apollo 11 spacecraft. Other Apollo Moon landings followed, and there were also landings of unmanned spacecraft by Russia and India, all on the bright side of the Moon visible from the Earth.

On 3 January 2019, China made history when their robotic spacecraft Chang'e-4 made a soft landing on the dark side of the Moon. It is the first spacecraft in history to attempt or achieve a landing on this unexplored area, which is never visible from Earth.

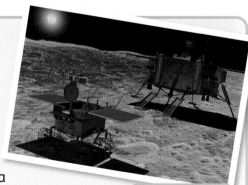
China's Yutu-2 exploration vehicle (front) and Chang'e-4 lunar probe (back) on the Moon

Exploring beyond the Earth

Think like a scientist!

How have scientists found about the Earth and beyond?

Humans first explored space simply by looking at the sky. During the day, they observed the apparent motion of the Sun. At night, they observed the apparent motion of the Moon, planets and stars. From these observations, they began to understand how the universe worked.

Then the **telescope** was invented. Improvements in its design over four centuries have allowed us to see more and more of the universe. These improvements have led to many new discoveries.

For the past 60 years, there has been another way to explore space – space flight. Space flight has allowed us to orbit the Earth, visit the Moon and launch telescopes into orbit. We can also send unmanned spacecraft, without any crew, to explore the Solar System.

telescope

1

a Choose one of these **astronomers**:

Zhang Heng

Ahmad ibn Muhammad ibn Kathir al-Farghani

Edwin Hubble

Jocelyn Bell Burnell

b Find out:
- when and where they lived
- the discoveries they made
- anything else about them that you think is interesting.

3

Do this activity as a class.

a Dress up as the astronomer you have chosen.

b Take turns to sit in the 'hot seat' and play the part of the astronomer you researched in Activity 1.

c The rest of the class will ask you questions, which you should try to answer.

2

a Do some research on the Ōyu Stone Circles in Japan.

b How did the ancient Japanese farmers use this giant sundial?

Scientific words
telescope
astronomers

Science in context

How space exploration keeps changing our understanding of the Solar System

The *Did you know?* box on page 133 gives examples of missions to the Moon.
You can read about missions to Mars below.

Mars missions by NASA

No humans have ever been to Mars, but scientists know a lot about its surface because spacecraft and rovers have been sent to explore the planet. Information collected by these vehicles are sent back to Earth as radio signals. Most of the earlier Mars missions were done by NASA (the National Aeronautics and Space Administration), the American space agency.

- In 1997, NASA's *Pathfinder* was the first spacecraft from Earth to land on Mars. It sent out a small rover to take photographs of the surface of the planet.

- In March 2004, *Spirit* and *Opportunity* landed on Mars, and stopped working only in 2010 and 2019. *Opportunity* had lasted 50 times longer than planned!

- *Curiosity* touched down on the surface of Mars in 2012. By 2020, it was still sending signals back to Earth.

New missions to Mars

In July 2020, during the Covid-19 pandemic lockdown, scientists from three countries still successfully launched new spacecraft to Mars while the Earth and Mars were at their closest – which happens only every 26 months:

- China's *Tianwen-I* mission started orbiting Mars in February 2021. Landing is planned for May/June 2021.

- The UAE spacecraft *Amal* (Hope), the Arab world's first mission to another planet, reached Mars in February 2021, orbiting without landing.

- NASA's *Perseverance* mission, the most sophisticated rover NASA has ever sent to Mars, landed on Mars on 18 February 2021. It also carries a helicopter named *Ingenuity*, the first aircraft that should fly sometime in 2021 in a controlled way on another planet.

An artist's painting of *Perseverance* and *Ingenuity* on the surface of Mars.

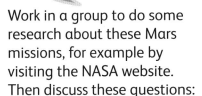

Let's talk

Work in a group to do some research about these Mars missions, for example by visiting the NASA website. Then discuss these questions:

a What did scientists hope to find out about Mars from these missions?

b What have they learnt from the results they have received so far?

c How have these results changed our understanding of the planet Mars?

d Do you think it is a good idea for different countries to do space exploration? Give reasons for your opinions.

Science in context

How satellites help us conserve water

To be able to produce enough food for the population of the planet, we need to make sure that we have enough clean water in the right places.

To help farmers, NASA has set up a system of Earth-observing satellites to help us make better use of the water we have on the planet. NASA scientists are looking at how they can use satellite technology in space to help create early warning systems, restore degraded waters, and improve the efficiency of water use.

Let's talk

a How many different ways can you think of that you use water every day?

b How much of this water do you think is wasted every day?

1

a Find out about the different ways that people could conserve water and use it better in their everyday lives.

b Create a poem, a rap or a song to let people know of the different ways they can conserve water, and why it is important for the planet.

Did you know?

97.5% of the water on Earth is salt water, leaving only 2.5% as fresh water.

Most of the fresh water is not available to us because 70% is frozen in the polar ice caps and 30% lies underground. This leaves less than 1% of the Earth's fresh water available for human use!

It is estimated that in some countries, people can use up to 575 litres of water a day, while people in developing countries only have about 19 litres to use every day.

What can you do to help conserve the Earth's fresh water?

What have you learnt about Earth and the Solar System?

Let's talk

Discuss these questions with a partner:

a What do you know now about the Earth and the Solar System that you did not know before?

b What is the most interesting thing you have learnt about Earth and the Solar System?

1

a Collect three balls of different sizes to represent the Sun, Earth and Moon.

b Use the three balls to explain to a partner the relative positions and movements of the Sun, Earth and Moon.

2

a Compared to the way the Sun appears from the Earth, how would it appear from Mercury? Explain your answer.

b Compared to the way the Sun appears from the Earth, how would it appear from Neptune? Explain your answer.

3

A Class 6 learner does not understand how we can see the same side of the Moon all the time, or why the shape of the Moon changes over time.

a Draw diagrams to explain these ideas to him.

b How could you test that he has understood your explanation?

What can you do?

You have learnt about the Earth and the Solar System. You can:

✔ describe the position and movement of the planets, the Moon and the Sun in relation to each other within the Solar System.

✔ review the names of the eight planets in our Solar System.

✔ explain why we only see one side of the Moon even though it is spinning.

✔ describe the changes in appearance of the Moon over its monthly cycle.

✔ describe how scientists can help us use water reserves on the Earth better.

Scientific dictionary

A

Absorbed Taken in

Accumulation The slow build-up of a substance

Angle of incidence The angle between the Normal and the incoming light ray

Angle of reflection The angle between the Normal and the reflected light ray

Angles The spaces between intersecting lines or surfaces, at the point where they meet; usually measured in degrees

Anticlockwise Turning in the opposite direction to the hands of a clock moving round

Aquaplaning Sliding uncontrollably on a wet surface

Arteries Blood vessels (tubes) that carry blood away from the heart

Artificial Made or produced by humans

Astronaut Space explorer sent up in a spaceship launched by a rocket to explore outer space

Astronomers Scientists who study the Solar System, galaxies and the universe

Axis The imaginary line around which an object spins, or the axis of a chart or graph

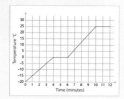

B

Bacteria Types of microbes; all bacteria are microscopic

Balanced diet Eating the right amount of a variety of foods from each food group

Balanced (forces) Forces that are the same strength but are acting in opposite directions; balanced forces do not change the movement of an object

Bedrock Solid rock that is normally buried beneath soil

Bioluminescence The ability of an organism to give off light

Blood The fluid in humans and other animals that delivers the essential materials for life to the body's cells

Blood vessels Tubes forming a network that carries blood around your body (part of the circulatory system)

Body Sensor Network A group of sensors that form a network and are attached to a patient's body in order to acquire data about their health

Boil When all of a material is changing from a liquid to a gas it is called boiling. All materials boil at different temperatures, the boiling point of water is 100 °C

Boiling A quick process that happens at the boiling point of a liquid when it changes into a gas. Boiling occurs throughout the whole of the liquid when its temperature reaches its boiling point

Boundary A line that shows the edge where one material ends and the next material begins

Brain One of the internal organs, located inside your skull; the body's control centre

Breathing The process of getting air into and out of the lungs

Breathing rate The number of complete breaths you take in one minute

Bronchi Two tubes inside your body that connect the trachea (windpipe) to the lungs

Bronchioles Tiny, hair-breadth tubes in the lungs branching off the bronchi

Burial The covering over of rock particles by more and more layers

C

Capillaries Tiny blood vessels that deliver water, oxygen and nutrients to the cells of the body, and carry away waste products

Cast fossils Fossils formed when plant or animal remains were covered quickly after death by sediments; in time, the sediments hardened to form rock and the bodies inside formed fossils

Cell An electrical component used in a circuit to generate an electric current. If more than one cell is used in a circuit, this is referred to as a battery

Cells The basic units that make up your body

Chemical reactions Changes in materials during which one or more new materials are formed

Circuit A complete loop, with no gaps, around which electricity can flow

Circuit diagram A diagram showing the components in a circuit and how they are connected to each other

Circuit symbols Symbols that represent components in circuit diagrams

Circulate Move around (the body) continuously

Circulatory system The body parts that work together to transport water, nutrients and oxygen to the cells of your body, and to carry away waste products that cells produce

Clay Soil type made up of very tiny soil particles that fit together closely; clay soil feels sticky

Complete breath The two-stage process of breathing in and breathing out

Components The separate parts that make up a complete circuit

Condensation The process in which a material changes state from gas to liquid (for example, water vapour in the air condenses to become water)

Conduct Allows electricity or heat to travel along or through it

Conservationists Scientists who work to protect habitats

Conserve Protect

Contact forces Forces that happen between two objects that are touching each other

Crust Outermost solid shell of a rocky planet

Cycles Series of events that are regularly repeated in the same order

D

Day Time taken for a planet to rotate once on its axis

Decay Break down, like the remains of dead animals and plants do to form humus

Defence mechanism An automatic reaction of the body against infectious diseases

Diaphragm A dome-shaped muscle underneath your lungs that moves up and down to let you breathe in and out

Diffuse Spread out over a large area

Disease When an infection causes damage to a person's organs or systems

Dissolve A solid that mixes in a liquid until it cannot be seen

E

Ecosystem All of the living and non-living things that are contained within a specific area, and how they interact and influence each other

Egg A cell produced in the reproductive organs of a female animal, sometimes called an ovum

Electrical conductivity How well a material conducts electricity

Ellipse A curved line forming a closed loop/ an oval shape

Energy The ability to do work

Energy flow diagram Diagram that shows where the energy from the Sun moves to in an ecosystem as animals feed

Equator An imaginary line around the Earth, halfway between the North Pole and the South Pole

Erosion The flow of water or wind that removes soil, rock or dissolved material from one place on the Earth's crust, and transports it to another place

Evaporation A slow process that happens at the surface of a liquid when it changes state from a liquid to a gas at any temperature below the liquid's boiling point

Excreted Ejected as waste

Exhalation Breathing out

Exhale Breathe out

External organs Organs on the outside of the body (such as the ears, nose and skin)

F

Fallopian tubes Parts of the female reproductive system

Feeding relationships These are represented by arrows and show how food moves from the plant or animal that has been eaten, into the organism that consumed it

Fertility rate The average number of children a woman gives birth to in her lifetime

Food chain A diagram that shows (from left to right in a row) how organisms in a particular habitat feed on one another. Arrows are used in a food chain to indicate the direction that energy flows through it

Food web A diagram, made up of several food chains linked together, that shows how the living things in a habitat rely on one another for food

Force diagram A diagram that uses arrows to show the size and direction of forces acting on objects

Force meter An instrument that is used to measure force in newtons (N)

Fossil fuels Natural fuels, such as coal, oil, petrol, diesel and gas, formed a long time ago from the remains of living organisms. The amount of fuel gets less when it is used

Fossils Remains of dead plants and animals from millions of years ago, or other traces left behind by them; these were covered quickly after death by sediments that with time hardened to form rock, turning the trapped bodies into fossils

Freezing The process in which a material changes state from liquid to solid

Freezing point The temperature at which a liquid freezes to become solid

Fungi Types of microbes; not all fungi are microscopic

G

Gas exchange A process by which oxygen and carbon dioxide move in opposite directions across a surface. In humans this happens in the lungs, as well as every cell in the body

Gas giants Very large planets made up mostly of gas; in our Solar System we have four gas giants: Jupiter, Saturn, Uranus and Neptune

Gestation period The time it takes for a foetus to develop, starting from fertilisation and ending at birth

Global warming A slow rise in the overall temperature of the Earth's atmosphere

Gravity The attraction between two masses; measured as weight of the force experienced by the mass being pulled down by gravity

Gravitational attraction The attraction that happens when there is an interaction between two masses. It is the amount of force a mass will experience, and is measured in newton per kilogram (N/kg). Gravitational attraction creates an object's weight

Greenhouse effect The trapping of the Sun's warmth in a planet's atmosphere

Greenhouse gases Gases that cause the greenhouse effect

H

Habitat The place where a plant, animal or other living thing lives

Heart One of the internal organs; the heart is a muscle in the chest with the function of pumping blood around the body

Heart rate A measure of how many times your heart beats in one minute

Humus A part of soil, formed by the decomposition of leaves, other plant material and animal matter by soil microorganisms

Hydroelectric power A way to make electricity by using water

Hygiene Ways we keep clean to help keep healthy and prevent disease

I

Igneous A type of rock formed from magma

Impermeable Does not allow liquid or gas to pass through

Infected When microbes have entered into a person

Infectious disease A disease that is capable of being transmitted to another host

Inflate To fill an object with air or gas

Inhalation Breathing in

Inhale Breathe in

Inherit Process by which the information in cells is passed from parents to offspring

Inner core The innermost part of the Earth

Inorganic A substance not made up of any organic material

Insoluble Does not dissolve in a liquid

Interact Have an effect on each other

Internal fertilisation The combining of an egg with a sperm during sexual reproduction inside the female body

Internal organs Organs inside the body (such as the brain, heart and lungs)

Internet of Things The use of the Internet to connect computing devices embedded in everyday objects, enabling them to send and receive data

Intestines Internal organs that form the lower part of the digestive system

Irreversible Cannot be reversed (changed back)

L

Lava Magma that reaches the surface, through a volcano or opening in the Earth's crust

LED lighting Energy-saving electronic devices that give off very bright light when an electric current flows through them

Life cycles The changes a living thing goes through, from the beginning of its life until its death

Life expectancy The average length of time that a person may expect to live

Light Light is emitted from a light source and is reflected off objects into our eyes so that we can see things

Light ray The pathway along which light travels from a light source

Light source An object that gives out light

Line graph A type of graph where the data is marked in dots (a series of points) joined by straight lines

Luminescent Materials that give off light

Luminous Objects that give out their own light

Lung capacity The volume of air that your lungs can hold

Lungs The internal organs that fill with air when you breathe

M

Magma Molten rock that is under the Earth's surface

Mains electric sockets The power points into which electric appliances can be plugged to connect them to the mains electricity supply in your home or school

Mantle The mostly solid bulk of the Earth's interior, made up of rock and magma

Mass A measure of the amount of matter an object contains, measured in kilograms (kg)

Materials The particular types of matter an object contains, measured in kilograms (kg)

Materials chemists Scientists who work to create new materials

Matter Makes up everything we can see around us; and is anything that has volume (takes up space) and mass

Melting The process in which a material changes state from solid to liquid

Melting point The temperature at which a solid melts to become a liquid

Menstruation Change in girls during puberty when they start having periods, usually monthly, of bleeding from the vagina as the blood-rich lining of the uterus is shed when a fertilised egg is not implanted (when pregnancy does not start)

Metamorphic A type of rock that has been changed by extreme heat and pressure

Metamorphism A process of changing rocks through heat and/or pressure

Microbes Germs that are too small to be seen with the naked eye

Microplastics Small pieces of plastic that are less than 5 mm in size, and are found in the environment because of plastic pollution

Microscopic An object that can only be seen with a microscope

Mirror An object designed to create reflections

Mirror image The image of an object reflected in a mirror

Molten Rock that is in a liquid state because of great heat

Movement Effects of forces applied to objects that cause them to change direction or position

N

Newton (N) The unit used to measure force

Non-contact forces Forces that happen between objects that are at a distance from each other

Nutrients Substances found in very small amounts in foods that animals need to stay healthy, and in the soil that plants need to grow

O

Opaque A material that light cannot travel through

Orbit The curved path taken by one object going around another

Organ Structure in the human body that performs a particular function (for example, the brain controls the body, and the heart pumps blood around the body)

Organic Material made of the remains of organisms such as plants and animals and their waste products

Organisms Living things

Outer core Layer of the Earth that lies above the inner core

Oxygen One of the gases that make up the air (animals take in oxygen from the air or water when they breathe, and oxygen is needed for burning)

P

Palaeontologists Scientists who study fossils in rocks

Parallel circuit A circuit in which components are connected in separate loops

Parasites Types of microbes; not all parasites are microscopic

Particles Tiny pieces, such as particles of soil

Periscope An instrument that uses mirrors to see over and around objects

Permeability How easy it is for a liquid to pass through a material

Phases (of the Moon) The movement of the Moon around the Earth makes it look as though the Moon is changing shape

Physical changes Changes that alter how a material looks or feels but do not produce new materials

Physical properties How a material looks or feels

Planets Rock or gas objects that orbit around a star are spherical in shape

Pollutants Materials such as plastic and oil that cause pollution and damage the environment

Polluted Not clean; containing pollutants

Pollution Damaging substances, especially chemicals or waste products, that harm the environment

Products Substances formed when a chemical reaction takes place

Puberty The stage of the human life cycle during which adolescents reach sexual maturity and become capable of reproduction

Pulse The regular throbbing of blood vessels as blood is pumped through them. Can be felt under the skin on some parts of the body (such as the wrists and neck)

R

Reactants Starting materials in a chemical reaction that interact with each other

Reflect Bounce off

Reflections Images of objects created by a very smooth surface

Refraction Change in direction of a wave at the boundary between transparent materials

Renewable A natural resource or source of energy that does not get less when it is used

Replacement level The fertility rate needed for a population to exactly replace itself from one generation to the next

Reproduction, sexual The production of a new living organism, created by combining information from parents of different types (sexes)

Respiratory system The body system with the function of breathing

Reversed Undone or changed back

Reversible Can be reversed (changed back)

Rocky planets Smaller planets made up of rock; in our Solar System we have four rocky planets: Mercury, Venus, Earth and Mars

Rotate Spin

Rotating Spinning

Rotation The action of spinning

S

Sand Larger, grainy soil particles with large spaces between them

Satellite An object in space that orbits a planet

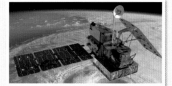

Savanna A tropical grassland

Sedimentary A type of rock formed from small rock particles deposited by air or water

Sedimentation The process of particles settling at the bottom of a liquid

Sediments Small rock particles broken down over time by weathering and that get carried away by erosion from one place to another, settling in layers

Series circuit A circuit in which components are connected one after the other in a single loop

Sexual maturity The age when humans can start reproducing, reached during puberty

Sexual reproduction The production of a new living organism, created by combining information from parents of different types (sexes)

Shadow An area of darkness created when light from a light source is blocked

Silt Very fine soil particles, usually from areas that were flooded before

Single-use plastics Objects made from plastic that are only used once before they are thrown away or recycled

Soil A mixture of tiny ground-down pieces of rock, dead plants, water and air

Soil erosion Washing away of topsoil by rainwater, made worse by the cutting down of trees

Solar panels A panel designed to absorb the Sun's rays as a source of energy for generating electricity or heating

Solar still A device that uses energy from the Sun to evaporate water, which then condenses and leaves clean water

Solar System Everything in orbit around the Sun

Solidification A process that happens when a molten substance turns into a solid

Soluble Dissolves in a liquid

Solution A mixture of a solid and a liquid that looks clear and has no particles floating in it

Solvent In a solution, the liquid into which a solid has dissolved

Sonoluminescence Light given off when sound waves pass through and burst bubbles

Species A specific type or group of animals or plants with the same characteristics, for example humans and dogs are different species

Specular (reflection) When rays of light hit a very smooth opaque surface, such as a calm lake, all the rays reflect at the same angle; this produces (forms) reflections and is called specular reflection

Sperm Cell produced in the reproductive organs of a male animal

Spherical Shaped like a sphere

Star A huge ball of very hot gas that produces a lot of light and heat; our Sun is a star

States of matter Matter can exist in different states; three important states of matter are solids, liquids and gases (matter can change from one state to another)

Steady speed When the speed of an object stays the same; does not increase or decrease

Stomach The internal organ where food is broken down; part of the digestive system

Subsoil The layer of soil under the topsoil on the surface of the ground

Support force A force that completely balances the weight of an object at rest

Surface area The total area of the surface of an object

Suspension A mixture of materials in which a solid is mixed with a liquid, but has not dissolved

T

Telescope An instrument that uses lenses and mirrors to allow people to see distant objects

Thermal conductivity How quickly heat conducts through a material

Top soil The top layer of soil

Toxic Poisonous

Trace fossils A fossil of a footprint, trail, burrow, or other trace of an animal rather than of the animal itself

Trachea An internal organ, also known as the windpipe; it is part of the respiratory system

Translucent Slightly see-through; a material that lets some light through, but you cannot see clearly through it

Transmission The spreading of microbes

Transparent See-through; a material that light can travel through, and you can see clearly through it

U

Unbalanced (forces) Forces acting in opposite directions that have different sizes; unbalanced forces change motion

Upthrust The force that pushes up on an object in water or air

Uterus Part of the female reproductive system where offspring gestate (live and grow) before birth

V

Vaporisation The process by which a material changes state from liquid to gas

Variable Factors that can affect the results of an experiment; all other variables, except the variable being tested, must be kept the same to ensure a fair test

Veins Blood vessels that carry blood from the capillaries back to the heart

Vertebrates Animals that have a backbone

Viruses Types of microbes; all viruses are microscopic

W

Waste Unwanted or unusable material or substance

Water vapour Water in the form of gas

Weathering Breaking the surface of a rock down into smaller pieces

Weight A downward, non-contact force that happens due to an attraction between two masses